Other Books by Deeds Publishing

War Stories – Utah Beach to Pleiku

Operation Iraqi Freedom I: A Year in the Sunni Triangle

"What Now, Lieutenant?"

You Don't Know Jack... or Jerry

"My Son is Alive..."

Freedom Express: Caught Between Iraq and a Hard Place

The Sooner the Better

Tales From the Mind

I Have Slipped the Surly Bonds of Earth and Danced the Skies

Deeds Publishing
Marietta, GA

www.deedspublishing.com

Wolfhound Reflections - A New Generation
Volume I

1st Battalion, 27th Infantry Regiment in Iraq

Operation Iraqi Freedom 07-09

December 2007 through February 2009

JB Jaso III

Deeds Publishing
Marietta, GA
www.deedspublishing.com

GRATEFUL ACKNOWLEDGMENT IS MADE TO THE COMBAT PHOTOGRAPHERS AND PUBLIC AFFAIRS SOLDIERS WHO CONTRIBUTED THE PICTURES AND STORIES THAT ARE IN THIS BOOK. ALL PHOTOGRAPHS ARE OFFICIAL US ARMY PHOTOGRAPHS.

PRINTED IN THE UNITED STATES OF AMERICA
PUBLISHED BY DEEDS PUBLISHING, MARIETTA, GA

FIRST EDITION, 2009

BOOKS ARE AVAILABLE IN QUANTITY FOR PROMOTIONAL OR PREMIUM USE. FOR INFORMATION WRITE DEEDS PUBLISHING, PO BOX 682212, MARIETTA, GA 30068 OR WWW.DEEDSPUBLISHING.COM.

ISBN 978-0-9776018-8-2

For All Past, Present, and Future Soldiers of the 27th Infantry Regiment "Wolfhounds" and their Families

From Wolfhound 6

Aloha from the Wolfhounds. The purpose of this book is to allow you, the reader, to review this command's most recent chapter in the 25th Infantry Division history. If you are neither a member of the Wolfhounds nor the Tropic Lightning Division, these reflections will provide you with a glimpse of the life our Warriors served while deployed under the 4th Infantry Division in Multi-National Division - Baghdad, Iraq.

As I look back over the past 15 months, I realize just how much each Soldier in our formation has accomplished. The Wolfhounds were responsible for two separate Operational Environments. As we first arrived in Iraq to conduct combat operations in OIF 07-09 we were assigned to the Taji Qada in Northwest Baghdad. This area was intensely rural and inundated with tribal complexities, shadowy insurgents, and limited infrastructure. Our mission was incredibly complex. In that area west of the Tigris River, the Wolfhounds lived among the people, operating from three Joint Security Stations in remote areas. On a daily basis, our Warriors engaged local leaders to work together to improve the quality of life for the Iraqis by focusing on neighborhood security watch with the Iraqi Forces.

Midway through the deployment, the Wolfhounds were called upon to shift focus and move east across the Tigris River to the Istaqlal Qada region. The battalion manned one Joint Security Station with five platoons and from three security stations patrolled an area the size of our previous area. One of our Company Teams, led by the B/52nd Anti-Tank separate company leadership served in Taji for the entire rotation.

Throughout the deployment, Charlie Company was detached from the Wolfhounds to be utilized as a Brigade asset and eventually attached to three different battalions within Second Brigade Combat Team - Warrior. They served with distinction in Sadr City and throughout all regions in MND-B.
When we moved across the river, Bravo Company was detached to go south to become a Brigade asset for Third Brigade, Fourth Infantry Division who assumed operational control of the Wolfhound Battalion in November 2008. Their dismounted patrols ensured Coalition Forces freedom of movement.

Throughout this 15 month deployment, the Wolfhounds performed close to 7,000 combat patrols. We refurbished 31 schools, providing for a better education and brighter future for the children of Iraq. In addition to the schools, we enabled local business owners to open numerous shops, boosting the economy in the local area. Among these shops were places such as grocery stores, photo shops, clothing shops, and barber shops, just to name a few. Working with Iraqi government leaders, we improved the sanitation, water quality and supply, and electrical network, greatly improving the lives of the local citizens. These areas are where Iraqis have begun to focus on their own self reliance in 2009.
As you look through this book and read the captions with each picture, you will get a feel for what we accomplished during our 15 months. Those 15 months were challenging, but extremely rewarding. We had some unfortunate moments as well, when we lost two of our Wolfhounds in combat. They will never be forgotten. To our Families, the 27th Infantry Historical Society, Peace Bridge, Friends and Sponsors of WOLFHOUNDS, we are humbled and honored to serve our nation's call and will now prepare for our next mission. I encourage you all to remain proud of each Wolfhound's service and sacrifice. These Warrior Army Families performed their duties with Discipline, an inspirational Attitude, and Teamwork that was phenomenal. All of this with NO FEAR!

RICHARD "FLIP" WILSON
LTC, IN
Commanding
1st Battalion, 27th Infantry Regiment
"WOLFHOUNDS"

LTC Richard "Flip" Wilson, a native of Brooklyn, NY, talks with a Taji city council member about detainees being released in the Taji Qada, northwest of Baghdad, April 17. (Photo by TSGT William Greer)

LTC Richard "Flip" Wilson, Commander, 1st Battalion, 27th Infantry Regiment, 2nd Stryker Brigade Combat Team, 25th Infantry Division, talks with key councilmen and sheiks during a conference held in Taji Qada, Iraq, Jan. 10, 2008. (Photo by TSGT William Greer)

A/V Team Tells the Wolfhound Story

In this day of communication, the need for information is at an all time high. Realizing this need, COL Stephen Banach, the 2/25 SBCT Commander from 2005-2007, decided that establishing Audio/Visual (A/V) teams at the brigade and battalion levels would assist during future operations in support of the Global War on Terrorism. COL Todd McCaffrey continued the plan after he assumed command of the brigade.

The 1-27 Infantry Regiment "Wolfhounds" quickly established an A/V Team that captured video and images of training events that were occurring in preparation for Operation Iraqi Freedom 07-09. The coverage provided by SSG James Hall, of Longview, TX; and SPC Brian Pierce, of Kingsford, MI, was instrumental for unit leaders to visually assess and improve their training.

Not only did the A/V team capture training, they participated in training of their own. They learned the techniques and ways to capture, edit, and produce a quality film.

Once in Iraq, the A/V team transitioned from capturing images of training to capturing the improvements being made in Iraq. "We focused on three target audiences," said Long Beach, CA native, SFC JB Jaso III, who was the "Wolfhound" Information Operations NCO and was responsible for the A/V Team. "First we targeted the Iraqi populace, then our Higher Headquarters, and finally the U.S. populace."

He said that the team had written nearly 200 articles for local newspapers, such as the Baghdad Now and Al Amal newspapers, notifying the community of improvements being made by the Government of Iraq (GOI) and security gains made by the Iraqi Security Forces (ISF).

"People were extremely excited to read about their communities and see the improvements being made," stated Pierce on the reactions when Iraqis received the newspapers.

"The GOI's forward momentum in coordinating their own essential services is phenomenal," continued Pierce. "It was great that they allowed us to assist in enabling them."

Once the community was notified, the team wanted to ensure that leaders at higher levels knew what the "Wolfhound" Soldiers were accomplishing, but most importantly the successes of the GOI and ISF in their area. The team produced well over 800 storyboards that were used at Multi-National Division – Baghdad, Multi-National Corps – Iraq, and Department of Defense briefs.

"It was extremely important to tell the "Wolfhound" story to our military leaders," Jaso stated. "The men did a tremendous job and if the A/V team didn't tell our leaders what we were doing, nobody would know the efforts that the men put in every day."

That same drive to tell the "Wolfhound" story carried over into notifying the U.S. populace of the accomplishments by producing nearly 1,000 media releases.

"It was important to show what we were doing over there," stated SGT Brad Willeford, of Yellow Springs, OH, the A/V Team Leader, on the importance of telling the "Wolfhound" story to the U.S. populace.

"The Soldiers liked seeing themselves in newspapers throughout the U.S.," Willeford continued. "It showed the Soldiers that we were not forgotten in the long 15-month deployment, and it was good for their friends and families to see what they were doing."

With well over 164,000 photos taken and nearly 1,000 patrols and operations conducted, the A/V Team was successful in telling the "Wolfhound" story. This book is a small sample of their work. ALOHA!

(From left to Right) SFC JB Jaso III, SPC Brian Pierce, SGT Brad Willeford

The Wolfhound Photographers

SSG JB Jaso III, of Long Beach, CA, captures images of Soldiers from HHC/1-27 Infantry Regiment "Wolfhounds," and community leaders following a sheik support council meeting in the Taji Qada, June 5. Jaso is the "Wolfhound's" information operations NCO.

SGT Brad Willeford, of Yellowspring, OH; SFC JB Jaso III, of Long Beach, CA; and SPC Brian Pierce, of Kingsford, MI captured images and stories of the 1-27 Infantry Regiment "Wolfhounds" 15-month deployment in the northern Baghdad region.

SPC Brian Pierce, of Kingsford, MI, records the grand opening festivities at the Bilal School in Baghdad's Adhamiyah security district on Oct. 11. Pierce is a infantryman with HHC/1-27 Infantry Regiment "Wolfhounds."

This Wolfhound gave his last full measure of devotion

SGT Jon Michael Schoolcraft III
27 December 1981—18 January 2008

SGT Jon Michael Schoolcraft III was born on December 27, 1981 in St. Mary's, OH. Jon entered the active service in June of 2001. Upon completion of One Station Unit Training at Fort Benning, GA, SGT Schoolcraft was assigned to A/1-18 Infantry Regiment, Conn Barracks Germany, where he deployed for his first time in support of OIF II.

In June 2005 he joined the Wolfhound Family and was assigned to C/1-27 Infantry Regiment, Schofield Barracks, Hawaii.

Jon's spirit, honesty, devotion to duty, and tireless pursuit of excellence inspired all who knew him.

Jon's interests and hobbies included diving, spear fishing, tattoos and of course he was a true Ohio State fan. Jon was an intelligent Soldier who would always look for some way to lighten up the mood and make people laugh.

SGT Schoolcraft's awards include the Bronze Star Medal, Purple Heart, Army Commendation Medal, Army Achievement Medal, Army Good Conduct Medal 3rd Oak Leaf Cluster, National Defense Service Medal, Iraqi Campaign Medal, Global War On Terrorism Expeditionary Medal, Global War On Terrorism Service Medal, Noncommissioned Officer Professional Development Ribbon, Army Service Ribbon, Overseas Service Ribbon, and the Combat Infantryman's Badge.

Sergeant Schoolcraft is survived by his wife Amber Leigh and his parents Jon and Cynthia Schoolcraft.

Schoolcraft was the kind of guy who just wanted to make everybody laugh. - PFC Cody

It didn't matter what the situation was, Michael Schoolcraft was the guy that would make you laugh and brighten up your mood. - PFC Riddle

When we got to Iraq, I remember Schoolcraft saying "I'm going to stay positive, because everyone else is depressed, I'm going to be positive." - CPL Young

A few days before the mission Mike expressed his feelings that he wanted to go back home to be with Amber. Well brother you're home now and you'll be missed by everyone in this platoon! I'll especially miss your text messages every weekend to go diving! Rest in peace brother, rest in peace. - SGT Walker

GONE BUT NOT FORGOTTEN.

A funny, outgoing man who valued his family and his brothers over anything. I'll miss you Schooley! - SPC Johnson

I'll miss cracking jokes, talking about life, and hitting up the beach every weekend. We'll miss you Schooley, see you when I get to heaven buddy. - Wood

Mike was the kind of guy who could make you laugh when it seemed like there was nothing to laugh about. - SGT Mathie

Every time I needed to know about new CD's that came out, Schoolcraft was here. - SPC "Medski"

There are no words to express how we feel, Schoolcraft you're in God's hands now. I am proud to have served with you, we love you and will miss you. - Harry J.

I'll remember you in my prayers at all times. Romans 1:9-10. - SPC Scales

Awesome friend and Soldier who would do anything he could for a friend. -SPC Lopez

I will never forget Michael. He always lightened up the mood, and I could always go to him for advice. A truly great American. - SPC Cochran

This Wolfhound gave his last full measure of devotion

SPC Gregory Brian Rundell
18 April 1986 – 26 March 2008

SPC Gregory Brian Rundell was born April 18 1986 in St. Paul, MN. He entered active service in August 2005.

Upon completion of One Station Unit Training at Ft. Benning, GA, SPC Rundell was assigned to B/1-27 Infantry Regiment, Schofield Barracks, HI. Gregory enjoyed hockey and football as a fan of both the Minnesota Wild and Minnesota Vikings. His hobbies included drawing, anything dealing with computers, and reading science fiction.

Gregory was a Soldier who loved to help his fellow Wolfhounds. If he had it, he would give it, no matter what it was. Gregory took great pride in his position as a vehicle commander with 1st Platoon Borzoi. After finishing his military service, his desire was to serve as a police officer.

SPC Rundell's awards include the Bronze Star Medal, Purple Heart, Army Good Conduct Medal, National Defense Service Medal, Iraqi Campaign Medal, Global War On Terrorism Service Medal, Army Service Ribbon, Overseas Service Ribbon, and the Combat Infantryman's Badge.

SPC Rundell is survived by his mother, Joanne Richardson; his Godfather Mike Niven; Brothers Kyle, Wayne, Andy and sisters Andrea and Desiree; and nephews Dane and Ben.

The best friend anyone could have! He was always a guy you could go to for anything. We will never forget you. - SPC Dustin Smith

He was the most generous person you ever met. A great Soldier doing an NCOs job extremely well!
- SSG Verrett

Would always drop what he was doing to help square away any Soldier.
- SGT Beer

YOU WILL BE MISSED.

Rundell, you'll be missed and never forgotten. Thank you for being a part of mine and my wife's lives, we will always remember you. - SPC Waldie

One of the best Soldiers I have ever had! He was like a son to me. He will never be forgotten. Words will never describe the love I had for him. - SSG Santos

I just want Rundell to know that we loved him and will miss him. - SGT Ortiz

G-Rundell, you will be sorely missed! We will never forget that "Joker" of a smile! "Wolfspirit"!
- SPC Stryker

SPC Rundell was a very giving man. Just a go to kind of guy. He will be missed.
- SPC Lindsey

Rundell was a good guy and a great friend. He was always there for me and I'm going to miss him. - SPC Klui

I remember when I would go to the gym and didn't have time to eat. There was always a Taco on my bed. This showed his thought and caring towards others. - SPC Rivera

Greg "Hips" Rundell was one of my best friends and roommate for the last 2 ½ years. He was a great person in mind and spirit. He always walked tall no matter what was happening. I love and miss him and will never forget him. - SPC Peptis

Took a lot of ribbing, but still managed to give a smile and an occasional punch in the nose. Your spirit will always be with us! -CPL Carney

Wounded Wolfhounds

PVT Zachary Ashbaugh (C/1-27 IN)

SSG Donald Fabian (TPT 1624)

SGT John Hanson (C/1-27 IN)

PFC Steven Riddle (C/1-27 IN)

SPC Esteban Rodriguez (C/1-27 IN)

SPC Gregory Rundell (B/1-27 IN)

SGT Jon Michael Schoolcraft III (C/1-27 IN)

SGT Abe Smith (C/1-27 IN)

PFC Bradley Thomas (C/1-27 IN)

PFC Jeremy Thomas (C/1-27 IN)

SGT Tomee Phetsisouk (B/1-27 IN)

PV2 Jaylon Ward (C/1-27 IN)

PV2 Toby Wilson (C/1-27 IN)

Wolfhound History 2007-2009

The 1-27 Infantry Regiment deployed to the Sunni Arab Region Northwest of Baghdad, Iraq as a 759 personnel task force from 08 December 2007 to 22 February 2009 in support of OPERATION IRAQI FREEDOM 07-09. During Task Force Wolfhounds' fifteen months of service, they conducted countless combat operations against Iraqi insurgents throughout two separate battlespaces covering a total of over 330 square mile area of responsibility (AOR). The Wolfhounds endured indirect fire and improvised explosive devices, yet never wavered or faltered. Throughout the entire deployment, the task force came under enemy fire several times.

The mission of the 1st Battalion, 27th Infantry Regiment was to conduct full-spectrum operations to control Operational Environment East in order to disintegrate Al-Qaeda in Iraq (AQI), Jaysh al-Mahdi (JAM) and Special Groups (SGs), protect the northern flank of the 3rd Brigade, 4th Infantry Division, partner with local Iraqi Security Forces, and develop local governance capabilities.

On 18 January 2008, just days after Transfer of Authority (TOA), Task Force Wolfhounds suffered their first casualties of the deployment. While C/1-27 IN was operating as Task Force Spear, a Stryker Brigade Combat Team - Warrior asset, the second Stryker of a three Stryker convoy was struck by an Improvised Explosive Device (IED) under the crew compartment. During this attack, C/1-27 IN sustained numerous casualties resulting in 1 Coalition Force Killed In Action (KIA) and 9 Coalition Force Wounded In Action (WIA).

On 14 February 2008, while providing overwatch for Route Raiders West, a squad from B/1-27 IN took small arms fire from at least one enemy rifleman. In this engagement, a Soldier from B/1-27 IN received a gunshot wound to the arm before the insurgent was run off. The valiant actions of this squad, as they defended their position and got the wounded Soldier to safety and medical treatment, were outstanding.

On 10 March 2008, while conducting a mounted combat patrol, 1-27 IN Mortars' convoy was engaged by the enemy with an IED. The IED attack resulted in a direct hit by two Explosively Formed Penetrators (EFPs) to the right side of the second Stryker in the convoy. As soon as the vehicle stopped, the squad dismounted, swept the area for secondary IEDs, and provided security. The Stryker was recovered and no KIA or WIA resulted from this attack.

PVT Daniel Perryman, a Spring Hill, FL, native, gives a boy a high-five during a patrol in the Istaqlal Qada, northeast of Baghdad, July 27, to assess the checkpoints in the area. Perryman serves as an infantryman with B/1-27 Infantry Regiment "Wolfhounds." (Photo by SGT Brad Willeford)

On 23 March 2008, while returning from an EOD response, the fourth vehicle in a C/1-27 IN convoy was struck by an IED. The IED penetrated the vehicle hull in several places and struck both tires on the passenger side of the Stryker. Immediately, they dismounted, swept the area for secondary IEDs and provided security until the recovery team arrived. No KIA or WIA were reported.

On 26 March 2008, a guard tower at Joint Security Station Sheik Amer received precision small arms fire from an unknown location. This sniper attack was believed to be related to the murder of a local Sons of Iraq (SOI) leader and sheik. This attack resulted in 1 Coalition Force KIA. The sniper was eventually reported KIA when he tried to escape after being captured by the SOI.

On 26 March 2008, C/1-27 IN deployed from Taji to Sadr City in order to help quell the violent uprising of the JAM and SGs operating in and around the Jamillah region. The need to occupy a position which tactically dominated the terrain became immediately apparent, and the company quickly established Patrol Base Texas in a four-story building in the center of Jamillah. Shortly after occupying the patrol base, Soldiers of C/1-27 IN began to take enemy sniper fire in their rooftop fighting positions from multiple directions - several times each day. This enemy sniper fire was extremely accurate, missing Soldiers by mere inches on many occasions, effectively suppressing the company's activities and ability to control the area.

A sniper squad arrived at Patrol Base Texas on 4 April 2008, and their presence made an immediate impact. They quickly assembled all of the pertinent information on the enemy's previous sniper engagements, conducted detailed time pattern analysis and assessments of impact angles, directions, and distances. Within hours they had created a plan to defeat the enemy, focused on identifying the enemy sniper hide positions and reacting rapidly once under attack. The squad presented their plan, recommended adjustments to friendly fire fighting position locations and also increased force protection measures to cover and conceal the new positions more effectively. In addition, they created several false sniper hide positions to draw enemy fire. This action achieved the desired result, and allowed our snipers to effectively identify enemy firing positions and suppress fire.

Over the next month, the squad of snipers went to work defeating the enemy sniper cell. They maintained constant over watch of known and suspected enemy firing and spotting positions, adjusting their own positions continually to meet the lightning and weather conditions at the patrol base. Whenever the patrol base came under attack, the squad fearlessly moved from position to position around the rooftop, exposing themselves to dialed-in enemy fire to identify muzzle flashes and movement from potential enemy sniper positions. The snipers were able to successfully identify, engage, and kill at least 11 enemy fighters throughout these engagements. With additional enemy suppressed, they destroyed the enemy's will to continue sniper attacks on the company which allowed the company to effectively patrol and maintain security without incident.

The sniper squad also maintained situational awareness and observation north of Route Gold, using their powerful optics to identify and kill armed enemy fighters at long distances, often at times when the enemy was engaging Iraqi Army Soldiers and fighting positions.

On 27 March 2008, C/1-27 IN was conducting patrols to disrupt the enemy in Jamillah in order to prevent IDF attacks on the International Zone in Baghdad. On one such patrol, the convoy was attacked using a complex ambush. The trail vehicle in the convoy was struck by an IED. The Stryker caught fire, and one Soldier was wounded in the attack. Third squad dismounted, moved to a rooftop and provided security. Second squad dismounted, swept around the blast site for secondary IEDs. The Weapons squad dismounted and swept for secondary IEDs. At the same time the Weapons squad was sweeping, the convoy came under small arms fire. Second squad and Weapons squad returned fire as they moved the wounded Soldier and placed him in another vehicle for evacuation. The Stryker was a total loss.

The Wolfhounds executed over three thousand six hundred combat patrols throughout both Area of Operations West and Area of Operations East during the deployment in order to defeat AQI, JAM and SGs, and their activities. These patrols proved extremely effective in facilitating the Task Force's ability to disrupt AIF routes into and through the Wolfhound Operational Environment (OE), degrade the AIF's ability to emplace IEDs, and prevent trafficking of weapons through the Task Force Area of Operations.

While operating in AO West, the Wolfhound's partnership with, and training of, the 700 Soldiers of the 36th Iraqi Army Brigade was second to none. The Task Force's dedication to the transition of the brigade to battlespace ownership consumed an enormous amount of time and effort. The Task Force regularly engaged the Brigade Commander and trained him on combat operations. The end result of these efforts created a well trained commander and staff that was able to assume battlespace in a very austere and dangerous part of the Taji Qada. Upon removal of 1-27 IN to another battlespace, the 36th IA was more than capable of assuming an even larger portion of the brigade's battlespace with positive results and accolades from the commander.

Upon the battalion's change of mission from the Taji Qada to east of the Tigris River, in August 2008, the Wolfhounds began work to enable both the Istaqlal Qada and Adamiyah security district representatives in the battalion OE. As a result the Task Force was able to influence the essential services through hard work and dedicated engagements that enabled the local Iraqi government to establish a larger presence with the provincial and rural councils. Task Force influence ensured that the local populace received the vital essential services needed and established a more defined, comprehensive governmental system. This system enabled both Iraqi and coalition force reconstruction operations to be deconflicted and focused on essential needs.

Again the Task Force taught, coached and mentored the 2-1 National Police Brigade to become a more lethal and viable source of security under it's tenure. They grew from a static security force to a maneuver brigade capable of executing the most complex of tasks. The Wolfhounds created an environment in their own battalion that was the forcing function for company commanders to fully partner with the Iraqi National Police battalions. This was met with outstanding results. Through partnership with the Al Zahour Iraqi Police

District, which included five local police stations, one patrol police station and one Emergency Response Unit, improved operational employment capability through establishment of a local training plan and combined patrols. The Wolfhounds trained a total of 375 Iraqis serving in the Iraqi Police Force and the Iraqi National Police Force in an effort to establish an Iraqi solution to the region's instability.

After completing the dissemination of Baghdad Now newspapers in the Hussaniyah town of the Istaqlal Qada on September 26, 2008, 2nd Platoon A/1-27 IN was struck by an EFP while exiting the city on Route Ninja. The EFP hit the front of the lead Stryker vehicle, causing small damage, and 1 Coalition Force WIA. Approximately one week later, during a Company (+) operation, the men responsible for emplacing the EFP were detained.

In November 2008, operational control of Task Force Wolfhounds (-) was given to 3rd Heavy Brigade Combat Team, 4th Infantry Division. As this occurred, B/1-27 IN (-) was sliced from the Task Force and sent to Forward Operating Base War Eagle where, in combination with Iraqi Security Forces, they conducted interdiction operations denying EFP cells influence and activity along Route Plutos from the Muthana Bridge to the Aeros-Plutos Intersection. B Co was very successful in preventing EFP emplacement and disrupting the EFP cell until the Brigade could develop enough intel to destroy the cell.

On Christmas Day, 2008, A/1-27 IN (-) was ordered South into the International Zone of Baghdad. There they conducted interdiction operations to deny the viable threat of Vehicle Borne IEDs during the holiday season. Their efforts ensured the International Zone was safe for Soldiers, civilians and Iraqis who either worked there or were on a much needed pass from their units.

On January 28, 2009, 2nd Platoon, A/1-27 IN was tasked to conduct a combined patrol with the Al Zahour Iraqi Police (IP) to verify the safety of polling center sfor the upcoming Iraqi Provincial elections. While traveling north on Route Dover to meet with the IP, a small IED exploded between the second and third vehicles, resulting in no injuries and vehicle damage. Members of the 6th Iraqi Emergency Response Unit and the 2nd Brigade, 1st Iraqi National Police Division quickly responded to the attack to assist the "Wolfhounds."

Despite the danger and the daily contact with insurgents, the Wolfhounds continually showed the discipline to balance combat operations with stability and support operations. The Wolfhounds worked tirelessly rebuilding the Taji Qada, Istaqlal Qada, and Northern Adhimiyah areas. They provided fresh water, increased medical services, built sewage systems, refurbished schools, paved roads, and expanded the reach of electricity. They established and trained local governments that focused on solving the issues of the people. They provided school supplies and clothing to the needy children.

Through continuous patrolling and local leader turn-ins, Task Force Wolfhounds was able to remove over 100 weapons caches from the battlefield. These caches contained munitions that would have been used against Coalition Forces ranging from 7.62 mm to 155 mm rounds and assorted types of IEDs. Also among the items found were various components used to make IEDs which would have been used to harm both Coalition Forces and Iraqi Forces alike. Additionally, the Wolfhounds uncovered numerous caches containing Unknown Bulk Explosives (UBE) seriously degrading the enemy's ability to attack Coalition Forces. Throughout the deployment, the Wolfhounds detained over sixty insurgents. These detainments had a direct impact on the security in the Wolfhound AO and allowed for the continuation of reconstruction efforts being implemented throughout the AO in order to rebuild the Iraqi infrastructure.

The Wolfhounds efforts both in combat operations and civic engagements neutralized insurgent activities in the region and allowed the Taji and Istaqlal areas, and the Iraqi people as a whole, to continue rebuilding their country. The discipline and professionalism displayed by Task Force Wolfhounds facilitated actions to defeat numerous indirect action attacks by IED, bombs, and rockets intended to disrupt operations. The tactical and technical competence while interacting with local sheiks assisted the local governments to reassert their control and authority, enabling a secure environment to initiate democracy.

The Task Force found 34 improvised explosive devices and spent over $28 million of Iraqi Commanders Emergency Response Program funds in reconstruction efforts to enhance the lives of the Iraqi people in the battalion battle space. Additionally, the Wolfhounds spent over $178K on micro grants to assist local businesses. Businesses such as grocery stores, barber shops, internet cafes, clothing stores and real estate have made a marked improvement in the local economies and the quality of life of the local populace. Task Force Wolfhounds has made a difference!

Soldiers from B/1-27 Infantry Regiment "Wolfhounds" are hit with a "dust devil" on a reflexive fire range at the National Training Center, Fort Irwin, CA, on Sept. 1, 2007. (Photo by PFC Stephen Miller)

SGT Pedro Ortiz, a Lajas, Puerto Rico native, gives the "Shaka" while preparing for training at the Southern California Logistics Airport in Barstow, CA, on Sept. 9, 2007. Ortiz is an infantryman with HHC/1-27 Infantry Regiment "Wolfhounds."
(Photo by CPT Chad Chenoweth)

PVT Justin Do, a Vancouver, WA native, does his best to stay hydrated in the hundred degree temperature while training at the National Training Center, Fort Irwin, CA on Sept. 1, 2007. Do is an infantryman with A/1-27 Infantry Regiment "Wolfhounds."
(Photo by PFC Stephen Miller)

CPL Jared Bruyneel, a San Bernardino, CA native, flexes in preparation for training at the National Training Center at Fort Irwin, CA on Sept. 1, 2007. Bruyneel is an infantry team leader with A/1-27 Infantry Regiment "Wolfhounds." (Photo by PFC Stephen Miller)

SPC Brian Pierce, a Kingsford, MI native, qualifies expert with his M4 carbine at the National Training Center, Fort Irwin, CA, on Sept. 3, 2007. Pierce is an infantryman with HHC/1-27 Infantry Regiment "Wolfhounds." (Photo by CPT Chad Chenoweth)

SGT James Wilson, a Pratville, AL native; and PFC Jose Carrillo, a Los Angeles native, conduct repairs to a vehicle that stopped because of maintenance problems at the National Training Center, Fort Irwin, CA, on Sept. 5, 2007. Wilson and Carrillo are wheeled vehicle mechanics for the 1-27 Infantry Regiment "Wolfhounds." (Photo by SSG JB Jaso III)

Hamilton, OH native, 1LT Chad Chenoweth and Brawley, CA native, 2LT Richard Pace discuss the operation of the live fire exercise that Pace's platoon will be executing on Sept. 6, 2007. Chenoweth is the battalion S3 Air officer, while Pace is a infantry platoon leader with C/1-27 Infantry Regiment "Wolfhounds." (Photo by SSG JB Jaso III)

Laredo, TX native, SSG Roberto Zuniga holds a sign asking for a ride to National Training Center's Forward Operating Base Seattle on Sept. 14, 2007. Zuniga is an infantry squad leader with A/1-27 Infantry Regiment "Wolfhounds." (Photo by PFC Stephen Miller)

MAJ Ross Coffey, a Seattle native, asks the question that was on every mind while at the National Training Center, Fort Irwin, CA, on Sep 13, 2007. Coffey served as the executive officer for the 1-27 Infantry Regiment "Wolfhounds", who was responsible for commanding "Wolfhound Yankee" command post. (Photo by CPT Chad Chenoweth)

Another day comes to a close before night operations begin with a beautiful Southern California sunset in the 1-27 Infantry Regiment "Wolfhounds," area of operations at the National Training Center, Fort Irwin, CA, on Sept 16, 2007. (Photo by SSG JB Jaso III)

CPT Zachary Reed, a Boston native, discusses plans for a company-level operation during the National Training Center rotation, on Sept. 24, 2007, preparing the unit for combat operations in Iraq. Reed is the fire support officer for the 1-27 Infantry Regiment "Wolfhounds." (Photo by CPT Chad Chenoweth)

2LT John Busuego, native of Ontario, CA, gives a thumbs up to his sleeping arrangements at the conclusion of two-months of training at the National Training Center, Fort Irwin, CA, on Oct. 1, 2007. Busuego is the chemical officer for the 1-27 Infantry Regiment "Wolfhounds." (Photo by SSG JB Jaso III)

The 1-27th Infantry Regiment "Wolfhounds" CRT while at Forward Operating Base Seattle, National Training Center at Ft Irwin, CA on Sept. 25, 2007. (Courtesy Photo)

SGT Mitchell Saltsman, of Bowling Green, KY, ground guides a disabled Stryker being towed by SPC Lee Coddington of Menomonie, WI onto a flatbed trailer for transport to San Diego, CA on Oct 2, 2007. (Photo by CW3 Danny K. Taylor)

Soldiers from HHC/1-27 Infantry Regiment "Wolfhounds" gather one last time with their family before departing Schofield Barracks, Hawaii for Iraq on Dec. 8, 2007. (Courtesy Photo)

SGT Daniel Durbin, of Kokomo, IN; SPC Jose Bautista, of Kansas City, KS; and PFC Bret Leback, of Manitowoc, WI, use their laptop computers while waiting for their flight to the Middle East from Hickam Air Force Base, Hawaii on Dec. 9, 2007. Durbin, Bautista and Leback are infantrymen with HHC/1-27 Infantry Regiment "Wolfhounds." (Photo by SSG JB Jaso III)

Charlotte, NC native, SGT Michael Pharr; and Fresno, CA, native, SSG Michael Chance relax and listen to music on their laptop computers at Hickam Air Force Base while waiting for their flight to the Middle East on Dec. 9, 2007. Chance is an infantryman, while Pharr is the fire support non-commissioned officer for the 1-27 Infantry Regiment "Wolfhounds." (Photo by SSG JB Jaso III)

Soldiers from HHC/1-27 Infantry Regiment "Wolfhounds" enjoy a game of cards during their layover in Ireland on their way to the Middle East on Dec. 10, 2007. (Courtesy Photo)

Inarajan, Guam, native, SSG Jesse Okiyama throws the *shaka* during the flight to Kuwait on Dec. 9, 2007. Okiyama is an infantryman assigned to HHC/1-27 Infantry Regiment "Wolfhounds." (Photo by PFC Nicholas Pata)

SFC Jesse Wyant, a native of Fillmore, NY, tries on a leprechaun's hat at the Shannon International Airport in Ireland, during a layover from Schofield Barracks to the Middle East on Dec. 10, 2007. Wyant is the platoon sergeant for the battalion operations shop assigned to the 1-27 Infantry Regiment "Wolfhounds." (Photo by PFC Nicholas Pata)

Soldiers from HHC/1-27 Infantry Regiment "Wolfhounds" fly from Hickam Air Force Base to the Middle East on Dec. 9, 2007. (Courtesy Photo)

Roswell, NM native, SPC Christopher Odell and Charlotte, NC native, SGT Michael Pharr play light saber wielding super heroes in the transient housing tents at Camp Beauring, Kuwait on Dec. 11, 2007. Odell is an infantryman, while Pharr is a fire support non-commissioned officer with HHC/1-27 Infantry Regiment "Wolfhounds." (Courtesy Photo)

Clarksville, IN native, SSG James Rigdon; Jacksonville, AZ native, Jeremie Olive; and Mililani, HI native, SSG Douglas Pierson, wait in line to be served dinner at a dining facility in Kuwait on Dec. 14, 2007. Rigdon, Olive and Pierson are human resources non-commissioned officers assigned to the 1-27 Infantry Regiment "Wolfhounds." (Photo by SSG JB Jaso III)

Soldiers from HHC/1-27 Infantry Regiment "Wolfhounds" await transportation to the M4 weapons qualification range in Kuwait on Dec. 15, 2007. (Photo by SGT Brad Willeford)

SPC Jermaine Anderson prepares to begin weapons qualification training in Kuwait on Dec. 15, 2007, in preparation for combat operations in Iraq. Anderson is the Standard Army Maintencance System operator for the 1-27 Infantry Regiment "Wolfhounds."
(Photo by SPC Brian Pierce)

Salina, KS native, SPC Mackenzie Diaz conducts small arms weapons qualification at a range in Kuwait on Dec. 15, 2007, preparing for future combat operations in Iraq. Diaz is an infantryman with C/1-27 Infantry Regiment "Wolfhounds."
(Photo by SSG JB Jaso III)

SPC William Groves, a Springfield, OR native, holds up his target after zeroing his M4 carbine at a weapons qualification range in Kuwait on Dec. 15, 2007. Groves is a combat medic with HHC/1-27 Infantry Regiment "Wolfhounds (Photo by SSG JB Jaso III)

Randolph, MN native, PFC Jacob Bruno points to the impact location of his rounds during a zeroing segment of a weapons qualification range in Kuwait on Dec. 15, 2007. Bruno is a combat medic assigned to HHC/1-27 Infantry Regiment "Wolfhounds." (Photo by SSG JB Jaso III)

Manitowoc, WI native, PFC Brett Leback points to his target on the firing line of a weapons qualification range in Kuwait on Dec. 15, 2007, while Gautier, MS native, SGT Christopher Nelson looks on. Leback is an infantryman, while Nelson is a human resources sergeant assigned to HHC/1-27 Infantry Regiment "Wolfhounds." (Photo by SSG JB Jaso III)

SPC Jose Bautista, of Kansas City, KS; and SSG Jesse Okiyama, an Inarajan, Guam native, inspect Bautista's marksmanship skills at an M4 qualification range in Kuwait on Dec 14, 2007. Bautista and Okiyama are infantryman assigned to HHC/1-27 Infantry Regiment "Wolfhounds." (Photo by SSG JB Jaso III)

Yellow Spring, OH native, SGT Brad Willeford receives a closer than expected haircut by Fresno, CA, native, SSG Michael Chance on Dec 15, 2007 at Camp Beauring, Kuwait. Willeford and Chance are infantrymen assigned to HHC/1-27 Infantry Regiment "Wolfhounds." (Photo by SSG JB Jaso III)

Brown Deer, WI native, SSG Mark Lewis conducts maintenance and cleaning on his M4 carbine rifle following a qualification range in Kuwait on Dec 16, 2007. Lewis is an intelligence sergeant for the 1-27 Infantry Regiment "Wolfhounds." (Photo by SSG JB Jaso III)

Fort Ann, NY native, MAJ Patrick Aspland; San Antonio native, CPT Arnulfo Martinez; Union, NJ native, CPT Mark Awad; and Boston native, CPT Zachary Reed prepare to depart Kuwait on Dec. 16, 2007 in order to prepare for the battalion's arrival in Iraq. Aspland, Martinez, Awad and Reed are assigned to HHC/1-27 Infantry Regiment "Wolfhounds." (Photo by SSG JB Jaso III)

Seattle native, MAJ Ross Coffey prepares for vehicle roll over training in Kuwait on Dec. 17, 2007. Coffey is the executive officer for the 1-27 Infantry Regiment "Wolfhounds."
(Photo by SSG JB Jaso III)

SPC Daniel Van Houten, a native of Pasadena, TX, enjoys an energy drink on Dec. 18, 2007 during explosive ordinance classes held in Kuwait prior to moving to combat in Iraq. Van Houten is an infantryman assigned to HHC/1-27 Infantry Regiment "Wolfhounds."
(Photo by SSG JB Jaso III)

CPT Mark Awad, native of Union, NJ, is buckled up and ready to receive vehicle rollover training in Kuwait on Dec. 17, 2007. Awad is the intelligence officer for the 1-27 Infantry Regiment "Wolfhounds."
(Photo by SSG JB Jaso III)

SGT Javier Snider, a San Diego native, trains soldiers on tactical movement techniques at Camp Beauring, Kuwait on Dec. 18, 2007 during close quarters marksmanship training. Snider is an infantry team leader with C/1-27 Infantry Regiment "Wolfhounds."
(Photo by SPC Brian Pierce)

SGT Nickolus Maxwell, a Hutchinson, KS native, readies his fire team for the combat in Iraq with realistic training at Camp Beauring, Kuwait on Dec. 18, 2007. Maxwell is an infantry team leader with C/1-27 Infantry Regiment "Wolfhounds." (Photo by SPC Brian Pierce)

Miles, TX native, PVT Jaylon Ward prepares to enter and clear a room during close quarter's battle training exercise in Kuwait on Dec. 18, 2007. Ward is an infantryman with C/1-27 Infantry Regiment "Wolfhounds." (Photo by SPC Brian Pierce)

Cromwell, CT native, SPC Robert Haswell inspects his weapon during training modules in Kuwait on Dec. 18, 2007. Haswell is an infantryman with C/1-27 Infantry Regiment "Wolfhounds." (Photo by SPC Brian Pierce)

SSG Michael Chance, a Fresno, CA native, reads a newspaper to keep up with current events in Kuwait while waiting for an improvised explosive device class on Dec. 18, 2007. Chance is an infantryman assigned to HHC/1-27 Infantry Regiment "Wolfhounds." (Photo by SSG JB Jaso III)

CPT Shawn Mitchell, a native of Cambridge Springs, PA; and 1LT Peder Miller, a native of Damascus, MD, wait for their ride to Camp Taji from Baghdad International Airport on Dec. 22, 2007. Mitchell is the adjutant, while Miller is the tactical intelligence officer for the 1-27 Infantry Regiment "Wolfhounds." (Photo by SSG JB Jaso III)

SGT Tony Macias, of Wauchula, FL; SPC Jerimiah Sullivan, of St. Joseph, MI; and SSG Mark O'Neill, of Pittsboro, IN, fly from Kuwait to Baghdad International Airport on Dec. 24, 2007. All are infantrymen with A/1-27th Infantry Regiment "Wolfhounds." (Photo by PFC Stephen Miller)

PV2 Stephen Miller, of Batesville, AR; SPC Scott Weller, of Auburn, NY; and SPC Johnathan Read, of Manila, AR await their helicopter flight from Baghdad International Airport to Camp Taji on Dec. 24, 2007. All are infantrymen with A/1-27th Infantry Regiment "Wolfhounds." (Courtesy Photo)

Staff members from the 1-27 Infantry Regiment "Wolfhounds" enter a Camp Taji dining facility to enjoy a Christmas dinner on Dec. 25, 2007. (Photo by CPT Chad Chenoweth)

1LT Christopher Hellie, of St. Paul, MN, celebrates on Christmas Day, 2007 with a great big smile at a Camp Taji dining facility. Hellie is the executive officer for HHC/1-27 Infantry Regiment "Wolfhounds." (Photo by SSG JB Jaso III)

MSG Timothy Jackson, of Groton, NY; and 1LT Peder Miller, of Damascus, MD, enjoy their Christmas Dinner at a Camp Taji dining facility on Dec. 25, 2007. Jackson is the operations sergeant major and Miller is the tactical intelligence officer for the 1-27 Infantry Regiment "Wolfhounds." (Photo by CPT Chad Chenoweth)

Cakes were prepared by cooks at a Camp Taji dining facility for soldiers deployed to enjoy on Christmas Day, 2007. (Photo by CPT Chad Chenoweth)

MAJ Patrick Aspland, of Fort Ann, NY; and SSG James Rigdon, of Clarksville, IN, discuss the manifest of men arriving to Camp Taji on Dec. 25, 2007. Aspland is the operations officer, while Rigdon is the personnel NCOIC for the 1-27 Infantry Regiment "Wolfhounds." (Photo by SSG JB Jaso III)

New City, NY native, PFC Nicholas Pata, and Griffin, GA native, SPC Ryan Floyd prepare to build beds for themselves at Camp Taji on Dec. 30, 2007. Pata is an infantryman, while Floyd is a forward observer with HHC/1-27 Infantry Regiment "Wolfhounds."
(Photo by SSG JB Jaso III)

CPT Brian Sweigart, a San Antonio native, rejoices as the helicopters that are carrying his men arrive at Camp Taji's airfield on Dec. 25, 2007. Sweigart commands A/1-27 Infantry Regiment "Wolfhounds." (Photo by SSG JB Jaso III)

PFC Justin Hollis, of Medina, OH; and SPC Richard Bobbitt, of Brenham, TX, carry a new mattress to their container housing unit at Camp Taji, on Dec. 30, 2007. Hollis and Bobbitt are mortarmen assigned to HHC/1-27 Infantry Regiment "Wolfhounds."
(Photo by SSG JB Jaso III)

SGT Christopher Taylor, of Fort Worth, TX; SSG Napoleon Gonzales, of El Paso, TX; and SFC Jerry Cox, of Brooklyn, NY receive equipment for the container housing units at Camp Taji on Dec. 30, 2007. Cox is the battalion supply NCOIC; Taylor is the B/1-27 IN supply sergeant; while Gonzales is an infantry squad leader in B/1-27 Infantry Regiment "Wolfhounds."
(Photo by SSG JB Jaso III)

Kingsford, MI native, SPC Brian Pierce celebrates the New Year as he films the festivities held at Camp Taji on Dec. 31, 2007. Pierce is an infantryman assigned to HHC/1-27 Infantry Regiment "Wolfhounds." (Photo by SSG JB Jaso III)

East Hartford, CT native, PFC Jose Algarin; and Wahiawa, HI native, PFC Ricky Reed celebrate as the clock hits midnight on Dec. 31, 2007 at Camp Taji. Algarin and Reed are personnel service clerks for the 1-27 Infantry Regiment "Wolfhounds."
(Photo by SSG JB Jaso III)

CPT Andrew Farina, of Franklin, NH, speaks to the soldiers of B/1-27 Infantry Regiment "Wolfhounds" prior to their movement to Joint Security Station Sheik Amer on Jan. 2, 2008. Farina commands B/1-27 IN. (Photo by 1LT Timothy Hall)

Yellow Springs, OH native, SGT Brad Willeford talks to New City, NY native, PFC Nicholas Pata about the inbound containers arriving at Camp Taji on Jan. 3, 2008. Willeford and Pata are infantrymen with HHC/1-27 Infantry Regiment "Wolfhounds." (Photo by SSG JB Jaso III)

CPT Chad Chenoweth, of Hamilton, OH, receives information pertinent to the Wolfhound's mission from the outgoing unit during the relief in place process on Jan. 3, 2008 at Camp Taji. Chenoweth is the assistant operations officer for the 1-27 Infantry Regiment "Wolfhounds." (Photo by SSG JB Jaso III)

SSG Herman Sims, a Detroit native, sends situation reports during a patrol in the New Taji Nahia on Jan 4, 2008. Sims serves on the military transition team assigned to the 1-27 Infantry Regiment "Wolfhounds." (Courtesy Photo)

Defuniak, FL native, SPC Christopher Fails receives a class on the command post of the future system used to track the battlespace on Jan 3, 2008, during the relief in place process required to assume control of the area of operations. Fails is an infantryman assigned to HHC/1-27 Infantry Regiment "Wolfhounds." (Photo by SSG JB Jaso III)

St. Petersburg, FL native, SFC Herman King works at his desk during a late night session to complete the work for the day on Jan 5, 2008 at Camp Taji. King is the communications non-commissioned officer assigned to the 1-27 Infantry Regiment "Wolfhounds." (Photo by CPT Joshua May)

Falls, PA native, CPT Joshua May works at his computer to resolve battalion communications issues on Jan 7, 2008 at Camp Taji. May is the communications officer for the 1-27 Infantry Regiment "Wolfhounds." (Courtesy Photo)

SPC Ulises Arellano, a Fontana, CA native, enjoys a quick snow fall at Camp Taji, located northwest of Baghdad on Jan 10, 2008. Arellano serves as a human resources specialist for the 1-27 Infantry Regiment "Wolfhounds." (Photo by CPT Chad Chenoweth)

Taji Area Changes Hands

By SPC Aaron L. Rosencrans
2nd Stryker Brigade Combat Team Public Affairs

CAMP TAJI, Iraq — The past week marked a significant change around the Taji area just north of Baghdad.

Taji has witnessed new units officially take charge of the area of operation, culminating Jan. 15, 2008 with the brigade's transfer of authority ceremony, which symbolized the official handing over of the area to the 2nd Stryker Brigade Combat Team, 25th Infantry Division from the 1st Brigade, 1st Cavalry Division.

Throughout the past 15 months, the 1st Bde. 1st Cav. Div. worked hard to create a secure environment for the local leaders and the government of Iraq to start rebuilding the essential services and the bonds between the leaders in the area.

Evidence of this progress was shown Jan. 10, when the local leaders had the third installment of the Northwest Baghdad Regional Security Summit at the Sheik Support Center.

Roundup, MT native, COL Paul E. Funk, II, commander of the 1st Bde. 1st Cav. Div. said when he and his brigade arrived in the area, it would have been impossible to hold a security summit where the local leaders could meet together and talk about security in the area and the future of Iraq.

Hudson, OH native, COL Todd B. McCaffrey, commander of the 2nd SBCT, 25th Inf. Div., congratulated Funk and his brigade for a job well done in the Taji area.

During McCaffrey's rotation in Iraq, he said he will continue to build a more secure environment for the government of Iraq and the local leaders to establish some stability in the region. In turn, the stable environent will allow for the Iraqi Police and the Iraqi Army to take over more of the operations that are currently supported by the coalition forces.

An example of McCaffrey's confidence in the security of the area, he went on a patrol through the Taji Market Jan. 14 and took the opportunity to talk to shop owners about their businesses.

Along with making the area stable, McCaffrey said he plans to bring more essential services, in an effort to improve the quality of life for the local Iraqi citizens. The extent to the growth of security and stability, he said, is up to the Iraqis.

Among the other units who had their transfer of authority ceremonies this week were the 1st Battalion, 14th Infantry Regiment; 2nd Bn., 14th Cav. Regt.; 1st Bn., 27th Inf. Regt.; and the Brigade Support Battalion.

Brooklyn, NY native, LTC Richard "Flip" Wilson, commander of 1-27 Infantry Regiment, and Las Vegas, NV native, CSM William Hain, command sergeant major of the 1-27 Infantry Regiment, uncase the battalion colors during a transfer of authority ceremony Jan. 13, 2008. (Photo by SPC Aaron L. Rosencrans)

A Soldier from B/1-27 Infantry Regiment "Wolfhounds" climbs into a hole where an illegal weapons cache was held in the East Anbar Province on Jan. 18, 2008.
(Photo by 1LT Timothy Hall)

SSG Bruce Hillier, of Twin Fall, ID, uncovers an illegal weapons cache in the East Anbar Province on Jan. 18, 2008. Hillier is an infantry squad leader with B/1-27 Infantry Regiment "Wolfhounds." (Photo by 1LT Timothy Hall)

Soldiers from B/1-27 Infantry Regiment "Wolfhounds" leave their mark after retrieving an illegal weapons cache in the East Anbar Province on Jan. 18, 2008.
(Photo by 1LT Timothy Hall)

Soldiers from B/1-27 Infantry Regiment "Wolfhounds," locate an illegal weapons cache in the East Anbar Province on Jan. 18, 2008. (Photo by 1LT Timothy Hall)

Fort Ann, NY native, MAJ Patrick Aspland; and Boston native, CPT Zachary Reed enjoy an evening meal that was attended by officers from the 36th Iraqi Army Brigade on Jan. 21, 2008. Aspland is the operations officer, while Reed serves as the fire support officer for the 1-27 Infantry Regiment "Wolfhounds." (Photo by SSG JB Jaso III)

MAJ Ross Coffey, a Seattle native, welcomes the officers of the 36th Iraqi Army Brigade to an evening festivity on Jan. 21, 2008 at a Camp Taji Dining Facility. Coffey is the executive officer for the 1-27 Infantry Regiment "Wolfhounds." (Photo by SSG JB Jaso III)

Cuyahoga, OH native, SPC Aaron Markwalder sits atop a World War-II era tank that was located at Camp Taji on Jan. 22, 2008. Markwalder is a network systems specialist assigned to HHC/1-27 Infantry Regiment "Wolfhounds." (Courtesy Photo)

SGT Nicolas Laduke, of Minneapolis, MN, troubleshoots the Common Remotely Operated Weapon Station (CROWS) that is mounted on the top of a High Mobility Multi-purpose Wheeled Vehicle (HMMWV) on Jan. 31, 2008. The CROWS allows a weapon to be safely fired from inside of the vehicle. Laduke is an armament and weapon repairer with the 1-27 Infantry Regiment "Wolfhounds" Combat Repair Team. (Photo by CW3 Danny K. Taylor)

JAMES

CASTANEDA

WASHINGTON

Iraqi citizens thwart suicide attack in al-Awad

2nd SBCT Public Affairs, 25th Inf. Div., MND-B

CAMP TAJI, Iraq — Two Concerned Local Citizens also known as Abna'a al-Iraq, were killed when a terrorist detonated a suicide vest outside a local photo shop in the market Feb. 5, 2008, in Al-Awad, northwest of Baghdad.

"This was a completely unsuccessful (terrorist) attack," said COL Todd B. McCaffrey, commander, 2nd Stryker Brigade Combat Team "Warrior." "The Sons of Iraq interdicted the bomber, preventing him from getting to his target. Their sacrifices exemplified the principles of courage and heroism."

The two Abna'a al-Iraq prevented the man from detonating the explosive vest filled with ball bearings in a crowd of local Iraqis. The selflessness of these men allowed time for surviving witnesses to recognize the threat and escape the blast.

The Iraqi Army Brigade Commander from the Al-Awad area is very committed to identifying the threat group responsible and preventing recurrences, McCaffrey said.

Multi-National Division – Baghdad Soldiers with the 2nd Stryker BCT, 25th Inf. Div. are working with the Iraqi Army, the Iraqi Police and the Abna'a al-Iraq to bring stability and security to the Taji region of Iraq located northwest of Baghdad.

Tommy Jackson, of Houston, TX; and SPC Donald Ross, of Albuquerque, NM, remove a 120mm mortar system from a Stryker Mortar Carrier Vehicle on Feb. 7, 2008. Jackson is a General Dynamics Land Systems contractor, while Ross is a wheeled vehicle mechanic. Both are attached to the 1-27 Infantry Regiment "Wolfhounds." (Photo by CW3 Danny K. Taylor)

Juan Alvarez , of Phoenix, AZ; and Carlito Deleon, of Tacoma, WA, replace a wheel drive assembly on a Stryker armored vehicle on Feb. 7, 2008. Alvarez and Deleon are General Dynamics Land Systems contractors attached to the 1-27 Infantry Regiment "Wolfhounds." (Photo by CW3 Danny K. Taylor)

SSG Mark O'Neill, of Pittsboro, IN; SGT Michael Rios, of Sebastian, TX; and SPC Collin Kelly, of Peachtree City, GA, prepare to conduct equipment inventory at Camp Taji on Feb. 10, 2008. All are infantryman with A/1-27 Infantry Regiment "Wolfhounds." (Photo by PFC Stephen Miller)

SGT Emelutero Lopez, of Detroit, MI, packs a wheel bearing with grease during trailer annual service on Feb. 13, 2008. The CRT performed scheduled services on all equipment assigned to the 1-27 Infantry Regiment "Wolfhounds" to ensure the unit could effectively move around their area of operation. Lopez is a wheeled vehicle mechanic attached to the "Wolfhounds." (Photo by CW3 Danny K. Taylor)

CPT John Goldsworthy, of Rosalia, WA, meets Sons of Iraq guards at Joint Security Station Hor al Bosh on Feb. 13, 2008, thanking them for their continued efforts in securing the New Taji Nahia. Goldsworthy commands B/52 Infantry Regiment, attached to the 1/27 Infantry Regiment "Wolfhounds." (Courtesy Photo)

1LT Timothy Hall, of Fairfax, VA; and CPT Andy Farina, of Franklin, NH, before Hall's last day as the B/1-27 Infantry Regiment "Wolfhounds," Executive Officer at Joint Security Station Sheik Amer on Feb. 14, 2008. Hall became the Battalion Logistics Officer upon leaving B Company; Farina commands B Company. (Courtesy Photo)

Stryker Soldier Serves Proudly in Iraq

By SSG JB Jaso III
1st Bn., 27th Inf. Regt., 2nd SBCT, 25th Inf. Div., MND-B

CAMP TAJI, Iraq – A long way from his Family, friends and hunting, PVT Benjamin Raulerson, a native of Jacksonville, AL, is here helping his unit, A/1-27 Infantry Regiment "Wolfhounds," 2nd Stryker Brigade Combat Team "Warrior," 25th Infantry Division, Multi-National Division – Baghdad, rid the Northwest Baghdad area of terrorism and criminal activity.

In his first month in Iraq, the 22-year-old infantryman has participated in more than 50 patrols, where he has assisted in identifying and seizing enemy cache sites and helped secure the area in order to provide the Government of Iraq the opportunity to provide essential services to the people.

"(It) feels good knowing that I'm assisting in removing items that can harm the U.S. and (Iraqi citizens)," said Raulerson, after returning from a patrol where his unit located a cache.

The cache had several mortars, rocket-propelled grenades and other items that could be used in improved-explosive devices.

Raulerson said his typical day includes patrols into various villages, where his unit assists the Iraqi Army in securing the area. What time he has left, he said he uses to relax.

The Soldier has many stories about his various patrols, but there is one incident that sticks out in his mind.

He recounted a time that while on patrol, he was invited to a local leader's home to have lunch. "I had no idea what I was eating, but it was good," he said. "I enjoy trying new things."

It is that adventurous spirit of trying new things that resulted in him joining the Army in April 2007.

After completing Basic Combat Training at Fort Benning, GA, he was assigned to the "Wolfhounds." He arrived at the unit just in time to join the other Soldiers at the National Training Center in Fort Irwin, CA, where he spent seven weeks honing his skills as an infantryman.

While at NTC, he was awarded the Army Achievement Medal for exemplary performance.

"(PVT) Raulerson is the best Soldier in my team," said CPL Douglas Urias, a native of Houston and Raulerson's team leader with A/1-27 Infantry Regiment, who added that he felt Raulerson may even be the best in the squad.

Raulerson said he attributes his initiative and natural leadership as a direct reflection of his father, who served in the Army as a military police officer in South Korea.

When he returns to Hawaii in approximately 14-months, Raulerson said he plans to explore the Hawaiian Islands, learn to surf and frequent the beaches. In the meantime however, he said he will continue to help protect the new freedoms of the Iraqi people.

PVT Benjamin Raulerson, of Jacksonville, AL, is assigned to A/1-27 Infantry Regiment "Wolfhounds." (Photo by SGT Brad Willeford)

(Left) Soldiers of the 2nd Stryker Brigade Combat Team, "Warrior" were awarded the Shoulder Sleeve Insignia-Former Wartime Service, or combat patch, northwest of Baghdad, Feb. 15, 2008. The "Tropic Lightning" patch symbolizes the unit's Hawaiian roots. The taro leaf is indicative of the origin of the 25th Inf. Div. in the Hawaiian Islands, while the lightning flash is representative of the manner in which the division performs its assignments.
(Photo by SFC Christina Bhatti)

(Right) Bryant, TX native 1LT Marco Montoya delivers newspapers to the Mass Media Center in Taji town, Jan. 13, 2008. Montoya is a platoon leader with C/1-27 Infantry Regiment "Wolfhounds."
(Photo by SFC Christina Bhatti)

(Below left) Round Rock, TX native, SPC Jeremiah Robles gives us the "shaka," on his way to the supply room at Camp Taji on Jan. 20, 2008. Robles is a infantryman assigned to B/1-27 Infantry Regiment "Wolfhounds."
(Photo by CPT Chad Chenoweth)

(Below right) SPC Daniel Blassingame, a native of Greenfield, OK, pulls security next to a Sons of Iraq guard at the Taji Market, northwest of Baghdad, Feb. 21, 2008. Blassingame is a mortarman with HHC/1-27 Infantry Regiment "Wolfhounds." The Wolfhounds regularly conducted patrols through the market to ensure the safety of the Iraqi people.
(Photo by SFC Christina Bhatti)

Stryker Soldier serves proudly in Iraq

By SSG JB Jaso III
1-27 Infantry Regiment "Wolfhounds"

Eight-months ago, PFC Steven Murphy, a native of Westford, MA, married his wife Sarah, in front of family and friends. Now he is thousands of miles away from his wife, and serves with B/1-27 Infantry Regiment "Wolfhounds" in support of Operation Iraqi Freedom.

"I was sick of not being a part of the war efforts," the 22-year-old infantryman said. He joined the Army in 2007 to deploy.

"It wasn't a matter of if, it was a matter of when," he said. He said he chose to become an infantryman because it "seemed more fun, and it was just right for me."

Before embarking on his 15-month deployment to Iraq, Murphy trained with his unit at the National Training Center, Fort Irwin, CA.

His training in California has been rewarded by his squad identifying enemy cache locations throughout his unit's area of operations in the Northwest Baghdad area. During his first cache recovery mission, his unit captured armor piercing rounds, which could have been used in attacks against either Iraqi Security Forces or Coalition Forces.

"We're doing good things over here, and we are noticing that [Iraqi's] are moving back in [the area] because it's safe," he said about his unit's mission to help secure an area northwest of Baghdad.

"Murphy is highly motivated and works incredibly hard," said 1LT Pete Walther, an Orlando, Florida native. Walther is Murphy's platoon leader, B/1-27 Infantry Regiment "Wolfhounds." "I've never seen a frown on his face."

Murphy's positive attitude and motivation has spread throughout his platoon, which is why he was selected as the only Soldier in his company to be apart of the division's combat patch ceremony.
"I'm going home with it on," he said.

Murphy will be home in Hawaii in approximately 13 months, but for now, he said he is happy he and his unit are making a difference in the lives of the people of Iraq.

PFC Steven Murphy a native of Westford, MA, and infantryman with B/1-27 Infantry Regiment "Wolfhounds" receives his 25th Infantry Division shoulder sleeve insignia-former wartime service, or combat patch, from COL Todd B. McCaffrey, commander, 2SBCT, 25th Inf. Div., MND-B, Feb. 15, 2008. (Photo by SPC Brian Pierce)

1LT Matthew Joyce, a Pittsburgh native, is promoted to captain by LTC Richard "Flip" Wilson at a ceremony on Feb. 24, 2008 at Camp Taji. Joyce is the mortar platoon leader assigned to HHC/1-27 Infantry Regiment "Wolfhounds." (Courtesy Photo)

1LT Timothy Hall, a native of Fairfax Station, VA, is promoted to captain by LTC. Richard "Flip" Wilson during a ceremony on Feb. 24, 2008 at Camp Taji. Hall is the battalion logistics officer for the 1-27 Infantry Regiment "Wolfhounds." (Courtesy Photo)

SPC Eric Barcomb, of Peoria, AZ, prepares to drive a Stryker armored vehicle on Feb. 19, 2008 in the East Anbar Province. Barcomb is an infantryman with B/1-27 Infantry Regiment "Wolfhounds." (Photo by 1LT Timothy Hall)

SSG Ahki Savea, of Pago Pago, Samoa patrols through the Al Awad region of the New Taji Nahia on Feb. 21, 2008, while inspecting Sons of Iraq checkpoints. Savea is an infantry squad leader with A/1-27 Infantry Regiment "Wolfhounds." (Photo by PFC Stephen Miller)

CPT Brian Sweigart, of San Antonio; and SSG Donald Fabian, of Miami, discuss economic development with Iraqi citizens in Taji, which is located northwest of Baghdad. Sweigart serves as the commander of A/1-27 Infantry Regiment "Wolfhounds." Fabian serves as a team leader with 310th Psychological Operations Company. (Photo by SGT Brad Willeford)

SPC Dustin Lapolla, a native of Providence, RI; and SPC Joshua Oden, of Lubbock, TX, traverse rocky landscape while on patrol in the Taji Qada, northwest of Baghdad. Both Soldiers are part of A/1-27 Infantry Regiment "Wolfhounds." (Photo by SGT Brad Willeford)

(from left) PFC Jeffrey Arnold, a native of Cheney, WA; SPC Dustin Lapolla, a native of Providence, RI; and SPC Benjamin Dang, a native of Orange Grove, NJ, secure all angles of a building during a patrol in the Taji Qada, which is northwest of Baghdad. All three Soldiers are from A/1-27 Infantry Regiment "Wolfhounds." (Photo by SGT Brad Willeford)

SPC Michael Miller, of Fort Worth, TX, inspects a house for traces of explosive material Feb. 26, 2008 during a search conducted in the Taji Qada, northwest of Baghdad. Miller serves as an infantryman with A/1-27 Infantry Regiment "Wolfhounds." (Photo by SGT Brad Willeford)

SSG Joseph Brown, a native of Oswego, NY, talks to a child while on patrol in the Taji Qada, which is northwest of Baghdad. Brown is an infantryman with A/1-27 Infantry Regiment "Wolfhounds." (Photo by SGT Brad Willeford)

SPC Johnathan Read, of Manila, AR, enjoys his downtime by playing Guitar Hero at Joint Security Station Al Awad on Feb. 29, 2008. Read is an infantryman with A/1-27 Infantry Regiment "Wolfhounds."
(Photo by PFC Stephen Miller)

PFC Stephen Miller, of Batesville, AR, communicates with his company and higher headquarters Feb. 26, 2008 in the Taji Qada, northwest of Baghdad. Miller is a platoon radio telephone operator from A/1-27 Infantry Regiment "Wolfhounds." (Photo by SGT Brad Willeford)

CPL Jared Bruyneel, of San Bernardino, CA, checks for explosive residue during a search of a suspected improvised-explosive device production facility Feb. 26, 2008 in the Taji Qada, northwest of Baghdad. Bruyneel is an infantryman with A/1-27 Infantry Regiment "Wolfhounds."
(Photo by SGT Brad Willeford)

PVT Justin Do, a native of Vancouver, WA, inspects the ground for improvised explosive devices and mines Feb. 26, 2008 in the Taji Qada, northwest of Baghdad. Do is assigned to A/1-27 Infantry Regiment "Wolfhounds." Along with the rest of his unit and the Iraqi Army, Do helps secure the Taji area. (Photo by SGT Brad Willeford)

Stryker Soldier serves proudly in Iraq

By SSG JB Jaso III
1st Bn., 27th Inf. Regt., 2nd SBCT, 25th Inf. Div., MND-B

CAMP TAJI, Iraq – The deployment to Iraq marks the first time PVT William Gentry has been out of the United States.

The Austin, Texas, native, said he was nervous and anxious to see what Iraq was like when he received his deployment notice.

"My heart was beating so fast that I thought it was going to come out of my chest," he said.

Gentry, an infantryman with B/1-27 Infantry Regiment "Wolfhounds," has been on more than 25 patrols, where he assisted in identifying and seizing enemy cache sites. His efforts, along with those of his fellow Soldiers, is helping to secure the area in order to provide the Government of Iraq the opportunity to provide essential services to the people of Iraq.

Now that Gentry has more than 25 patrols under his belt, he said he feels more at ease, which he says allows him to conduct his missions more thoroughly.

Sixty miles of walking per week takes it toll on the body, but despite his aching back and feet, Gentry said he sees his pain as a necessity to life.

"The way I see it," he said "life without suffering is impossible; those who live running from pain live no life at all. This job may seem easy to some but, for most, it is a test to find out how strong of a person we really are."

The quest to find his strength, and the desire to travel, he said, motivated him to join the Army in October 2006. Prior to joining the Army, Gentry had never been out of the state of Texas.

He currently lives with U.S. Soldiers and Iraqi Army Soldiers on a joint security station, which provides Gentry and his unit the ability to rapidly respond to activities that occur in their area. Although he lives on a smaller compound, he said he is thankful he can still communicate with those most important to him – his mom and four siblings.

When he returns to Hawaii in approximately 13 months, Gentry said he plans to go sky-diving and explore more of the Hawaiian Island's snorkeling. In the meantime however, he will learn all he can from the deployment and strives to stay positive and continue to help make life better for Iraqis.

(Top) PVT William Gentry, a native of Austin, TX, stands watch during a patrol with B/1-27 Infantry Regiment. The unit patrols an area northwest of Baghdad to help ensure security in the area and to provide the Government of Iraq the opportunity to work safely in improving the lives of the people.

(Bottom) PVT William Gentry, a native of Austin, TX, and an infantryman with B/1-27 Infantry Regiment, enjoys interacting with Iraqi children while on patrol in the Taji Qada, which lies northwest of Baghdad. He said "kids are fascinated by him and he loves it" when children approach him.
(Photos by SGT Brad Willeford)

SSG Kellen Hansen, native of Mar Vista, CA, speaks with children in a New Taji Nahia village on Feb. 26, 2008. Hansen is a reconnaissance scout team leader with HHC/1-27 Infantry Regiment "Wolfhounds." (Photo by CPT Chad Chenoweth)

SSG James Hall, native of Castilion, TN, speaks with a New Taji Nahia resident about possible illegal weapon caches hidden in this northwest Baghdad region on Feb. 27, 2008. Hall is a reconnaissance scout team leader with HHC/1-27 Infantry Regiment "Wolfhounds." (Photo by CPT Chad Chenoweth)

SPC Ronald Jarrard, a native of Holy Hill, FL, walks through the barren areas of the New Taji Nahia, while searching for weapons caches on Feb. 27, 2008. Jarrad is a reconnaissance scout with HHC/1-27 Infantry Regiment "Wolfhounds." (Photo by CPT Chad Chenoweth)

SPC Osvaldo Ortiz, a Rincon, Puerto Rico native, uses a metal detector while searching for enemy weapons caches in the New Taji Nahia on Feb. 27, 2008. Ortiz is a reconnaissance scout with HHC/1-27 Infantry Regiment "Wolfhounds." (Photo by CPT Chad Chenoweth)

Sturgis, MI, native, SGT Mark Gunthorpe reenlists for an additional four years on Feb. 12, 2008 at Camp Taji. Gunthorpe is an Infantryman with C/1-27 Infantry Regiment "Wolfhounds." (Courtesy Photo)

SSG Kellen Hansen, a native of Mar Vista, CA, holds an Iraqi girl who ran up to him during a patrol March 4 in the Taji Qada, northwest of Baghdad. Hansen is a Reconnaissance Scout assigned to HHC/1-27 Infantry Regiment "Wolfhounds." (Photo by SGT Brad Willeford)

CPT Chad Chenoweth, a native of Hamilton, OH, talks with local leaders and conducts assessments March 4 in the Taji Qada, northwest of Baghdad. Chenoweth is a Reconnaissance platoon leader assigned to HHC/1-27 Infantry Regiment "Wolfhounds." (Photo by SGT Brad Willeford)

SSG James Hall, a native of Longview, TX, provides security March 4 in the Taji Qada, northwest of Baghdad. Hall is a Reconnaissance Scout assigned to HHC/1-27 Infantry Regiment "Wolfhounds." (Photo by SGT Brad Willeford)

CPT Andrew Farina, a Franklin, NH native, discusses security plans with a local leader during the unit's security conference March 5, at Camp Taji. Farina's company has been working with Iraqi leaders to secure and improve the Taji Qada, northwest of Baghdad. Farina serves as the commander of B/1-27 Infantry Regiment "Wolfhounds." (Photo by SGT Brad Willeford)

(Right) SPC Christopher Odell, a Roswell, NM, native, serves lunch to Iraqi leaders during a security conference March 5, at Camp Taji. Odell serves as an infantryman from B/1-27 Infantry Regiment "Wolfhounds." (Photo by SGT Brad Willeford)

Iraqi Army HMMWV contact truck familiarization training with SFC Mark Clark, of Canon City, CO; and SGT Christopher Delaney, of Belleview, FL. The CRT trained the IA mechanics on familiarization and techniques to use with their new maintenance support contact trucks on March 8. Clark and Delaney are wheeled vehicle mechanics with the 1-27 Infantry Regiment "Wolfhounds." (Photo by CW3 Danny K. Taylor)

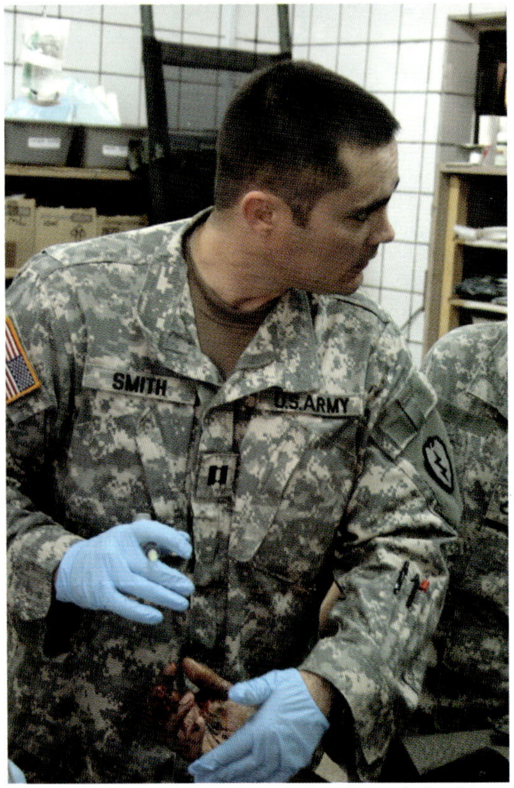

CPT Kenneth Smith, a native of Phoenix, and physician's assistant for 1-27 Infantry Regiment "Wolfhounds," treats a Son of Iraq (Abna al Iraq) volunteer who was injured during a improvised explosive device attack March 7.
(Photo by SPC Brian Pierce)

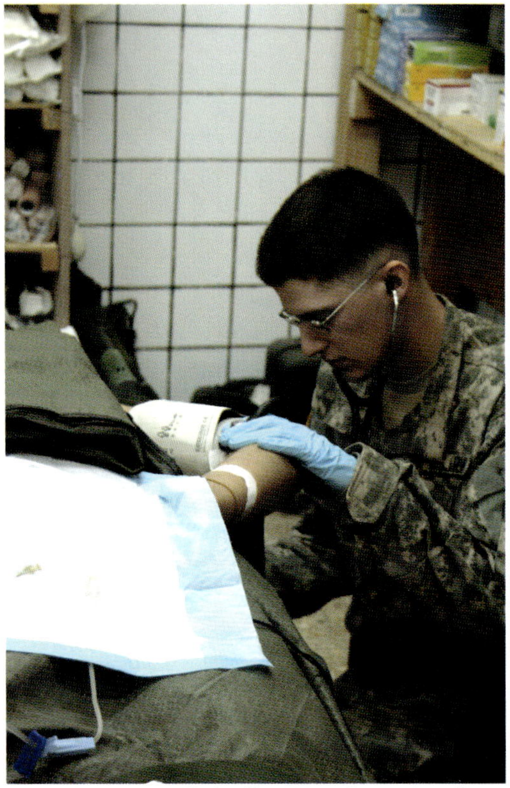

PVT Eric Young, a Clarksville, TN, native, and medic assigned to HHC/1-27 Infantry Regiment "Wolfhounds" checks the vitals of a Sons of Iraq (Abna al Iraq) volunteer March 7. The SoI volunteer was injured when an improvised explosive device exploded at the checkpoint he was manning.
(Photo by SPC Brian Pierce)

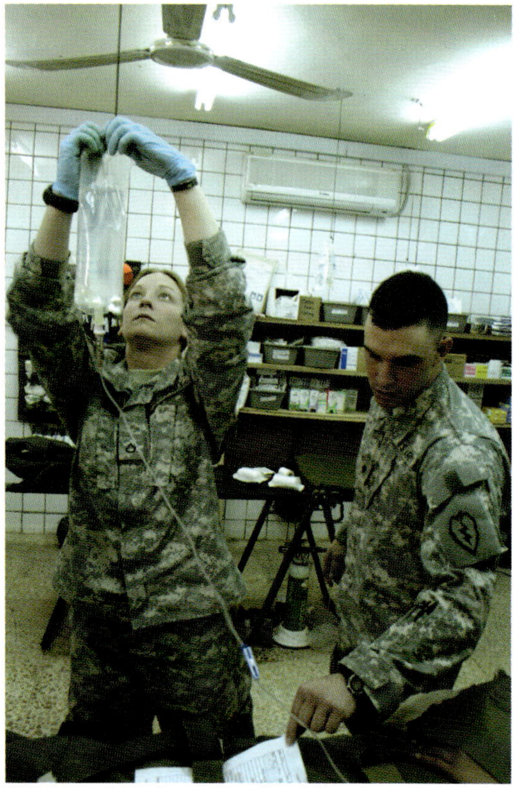

PFC Melanie Shelton, a Henniker, NH native and medic begins immediate medical attention on a Sons of Iraq (Abna al Iraq) volunteer who was wounded during an improvised explosive device explosion on March 7. SPC Ryan Maxfield, a Howell, MI native, and medic assigned to HHC/1-27 Infantry Regiment assists Shelton. (Photo by SPC Brian Pierce)

Falls, PA native, CPT Joshua May spent his personal time to build an area for Soldiers to relax (which he named "Carter's Cabana" after his son) near the container housing units at Camp Taji on Mar. 12. May is the communications officer for the 1-27 Infantry Regiment "Wolfhounds." (Courtesy Photo)

On Mar. 13, Soldiers of HHC/1-27 Infantry Regiment "Wolfhounds" spend the evening together at "Carter's Cabana," which Falls, PA native, CPT Joshua May built during his personal time. (Courtesy Photo)

CPL Nathan Adams, a native of Boise, ID, assists in making identification cards for Sons of Iraq (Abna al-Iraq) volunteers March 13 in the Taji Qada, northwest of Baghdad. Adams serves as a reconnaissance scout with HHC/1-27 Infantry Regiment "Wolfhounds." (Photo by SGT Brad Willeford)

SPC Collin Kelly, of Peachtree City, GA, plays with children on a merry-go-round at an Al Awad school on Mar. 17. Kelly is an infantryman with A/1-27 Infantry Regiment "Wolfhounds." (Photo by PFC Stephen Miller)

Stryker Soldier Writes New Chapter in Family History

By SSG JB Jaso III
1-27 Infantry Regiment "Wolfhounds"

As some 19 year-old males are going to college, hanging out with friends and beginning their adult life, one Multi-National Division – Baghdad Soldier is beginning his adulthood assisting in securing Iraq in support of Operation Iraqi Freedom.

PVT Derk Hayes, a Peru, IN native, is the youngest Soldier in his unit, C/1-27 Infantry Regiment "Wolfhounds."

From a very young age, Hayes knew he wanted to be a Soldier in the Army. "He always said that he was going into the Army," reflects Hayes' mother, Candace Hayes. She remembers him saying that when he was six or seven years-old.

By following through on that dream, Hayes opens a new chapter in the line of his family's service.

Hayes' great-uncle, Virgil Hayes, served in the Army during World War II and passed away in 1995, when Hayes was 6 years old.

"The last time I saw my Uncle Virgil, I told him I would join the Army," said Hayes on why he joined the Army in June 2007.

Not only did his great-uncle serve in World War II, but Hayes' grandfathers served in the military as well. His grandfather, Lyle Hayes, is a Pearl Harbor veteran who was awarded the Purple Heart due to injuries sustained in Japan's infamous attack on Dec. 7, 1941. His other grandfather, Hoot Gibson, retired after serving 30 years in the Air Force.

After attending Basic Combat Training at Ft. Benning, GA, Hayes was assigned to the "Wolfhounds" at Schofield Barracks, Hawaii, where he spent three weeks before deploying to Iraq.

"I knew it was going to happen, but not that soon," Hayes said. "Telling my mom was hard. I care about her more than anything."

His mom was at work when she received his text message and phone call notifying her that he was deploying in weeks.

"I broke down and just started to cry," Candace said, upon receiving the news from her son. "I just prayed, prayed and prayed; I was so scared." "It was so hard to prepare for (the deployment)," said Candace, almost four-months into Hayes' deployment. "I'm doing better and praying for him every day. Some days I cry, but he reassures me that he's fine."

Hayes said he communicates with his mom as much as possible, but when he can't call home, Candace said she talks with other military mothers in a social network. She is a member of the Web site www.milclub.net, which is a place for troops and their families to communicate and share their experiences. She communicates with other mothers who have children deployed and said "they are helping me get through this."

In his first four months in Iraq, Hayes has been on more than thirty patrols, where he assists with the security of the brigade's embedded Provincial Reconstruction Team as it travels throughout the Taji Qada, northwest Baghdad.

Hayes was presented with a 25th Infantry Division combat patch March 9 at a ceremony on Camp Taji. His service in Iraq is rewarding, and this ceremony was a "time to pause and recollect on the gains that we are making," said LTC Richard "Flip" Wilson, battalion commander for 1-27 Infantry Regiment "Wolfhounds."

After being presented with his combat patch, Wilson requested Hayes to join him in front of the company formation, where Wilson presented

Hayes a commander's coin.

"I felt great. Getting a coin inspired me to do better," Hayes said. "I've never seen (a coin) before, and I was really amazed. I'm keeping it in my pocket below my heart."

Getting better is what his mother said the military has helped her son with.

"The military changed him for the better," Candace said, after hearing that he was recognized by the battalion commander. "I'm really, really proud of him."

PVT Derk Hayes, a Peru, IN, native, who serves with C/1-27 Infantry Regiment "Wolfhounds", is presented a commander's coin for excellence March 9 by LTC Richard "Flip" Wilson, a Brooklyn, NY, native, who serves as the battalion commander of 1-27 Infantry Regiment "Wolfhounds." (Photo by SGT Brad Willeford)

(Right) PVT Derk Hayes, a native of Peru, IN, native, is assigned to C/1-27 Infantry Regiment "Wolfhounds." (Photo by SGT Brad Willeford)

NY firefighter leaves NY to Fight Fire in Iraq

By SSG JB Jaso III
1-27 Infantry Regiment "Wolfhounds"

On September 11, 2001, PFC Nicholas Pata, then a volunteer firefighter in Rockland County, NY, assisted rescue efforts after terrorists attacked the World Trade Center in Manhattan.

Twenty-five year-old Pata, a New City, NY native, no longer fights fires, he now fights terrorism as a radio-telephone operator assigned to HHC/1-27 Infantry Regiment "Wolfhounds."

"After 9-11, seeing and losing (many) friends that were firefighters, I felt I owed it to them to jump into the fight," Pata said. "The time I spent at Ground Zero made up my mind (to join)."

Pata joined the Army in January 2007, where he completed Basic Combat Training at Fort Benning, GA, and then was assigned to Schofield Barracks, Hawaii, and joined the "Wolfhounds" in June.

Less than two months later, Pata departed Hawaii with his unit to conduct training at the National Training Center in Fort Irwin, CA There, he used his medical skills gained as a fireman & emergency medical technician to save the life of a fellow Soldier who was suffering from a severe heat injury. As a result of his actions and performance during the training rotation, he was awarded the Army Achievement Medal.

After completing training in California, Pata took pre-deployment leave to relax before a 15-month deployment to Iraq. His leave was not all relaxation. He suited up and went back to work as a firefighter. Pata answered the last alarm minutes before he had to return from leave.

"Nick told (his fellow firefighters) before he left for Iraq to keep his 'bunk warm,' and (not to) 'mess with my gear; leave it right where it is,'" said Greg Tobin, a fellow Rockland County volunteer firefighter. "Ever since, he has been gone his gear is exactly like he left it. No one (has) dared to touch it – not out of fear, rather out of respect for the man who wore it."

When Pata left in October, he not only had to say goodbye to his biological Family, but he had to say goodbye to his other Family, his fellow firefighters, said Tobin.

"His passion as a firefighter to help his community is what Nick lives for. He is a unique person that has risen to the position of captain in the fire department. He feared no fire; he was always the first into a fire and the last one out, and always making sure he watched over the men he led. He is a brave man, very respected, and very missed by us at home. The community will be safer again when he comes home," he added.

While in Iraq, Pata assists his unit and the Iraqi Security Forces keep the Taji Qada, northwest of Baghdad, safe.

It is his experience under fire that has helped him be a calm presence here. "Pata always maintains his composure under pressure regardless of the situation," said MAJ Patrick Aspland, a native of Fort Ann, NY. Aspland is the executive officer for 1-27 Infantry Regiment "Wolfhounds." "His maturity and experience gives him this mannerism that he has things under control."

Pata was recognized by GEN David Petraeus, the commander of Multi-National Forces – Iraq, March 12 for his exemplary performance in Iraq. "Your great work is bringing new hope for the Iraqi people," stated Petraeus, during the award ceremony. "Keep up the terrific work!"

PFC Pata has been doing terrific work as he "is one of the best RTOs I've seen in my career; he is able to handle multiple tasks simultaneously," said MSG Timothy Jackson, a native of Dryden, NY, who serves as the operations sergeant major for 1-27 Infantry Regiment "Wolfhounds."

Pata has approximately one year left in Iraq before going back to fight a different kind of fire. Pata said he looks forward to going back to Hawaii and eventually returning to New York to continue to serve the people there.

PFC Nicholas Pata, a New City, NY native, who serves as a radio-telephone operator for HHC/1-27 Infantry Regiment "Wolfhounds," is presented a Commander's Coin for Excellence March 12 by GEN David Petraeus the commander of Multi-National Forces – Iraq. (Photo by SSG Christian Foster)

(Right) PFC Nicholas Pata, a New City, NY native, who serves as a radio-telephone operator for HHC/1-27 Infantry Regiment "Wolfhounds," was a volunteer firefighter in the Rockland County Fire Department prior to joining the Army. Pata assisted in the rescuing efforts after the attacks on Sept. 11, 2001.
(Photo courtesy of the Pata Family)

SGT Deante Long, of Columbus, OH, welds two steel plates together during a fabrication project on Mar. 18. The CRT often conducted welding and fabrication of necessary brackets and frames for use on various types of military equipment. Long is a wheeled vehicle mechanic with the 1-27 Infantry Regiment "Wolfhounds." (Photo by CW3 Danny K. Taylor)

PVT Joshua Echeverra, a native of Boise, ID; and PVT Justin Do, a native of Vancouver, WA, patrol the streets of the Taji Qada, northwest of Baghdad. The Soldiers are infantrymen assigned to A/1-27 Infantry Regiment "Wolfhounds." (Photo by SGT Brad Willeford)

Falls, PA, native, CPT Joshua May rides an all terrain vehicle on Camp Taji, moving tables and much needed items to and from the unit's tactical operations center on Mar. 20. May is the communications officer for the 1-27 Infantry Regiment "Wolfhounds." (Courtesy Photo)

SSG Donald Fabian, a Miami native, who serves as a team leader with the 310th Psychological Operations Company; and 1LT Steven Robinson, a Herndon, VA native, who serves with B/1-27 Infantry Regiment "Wolfhounds", talk with Iraqi residents about what they need to better their community March 20.
(Photo by SGT Brad Willeford)

SSG Bruce Hillier, a Twin Falls, Idaho, native who serves with B/1-27 Infantry Regiment "Wolfhounds", talks with several Iraqi girls while on patrol northwest of Bagdad March 20.
(Photo by SGT Brad Willeford)

SGT Bruce Kihleng, a native of Kosrae, Micronesia, diligently works to establish a network for the headquarters computers on Mar. 28. Kihleng is a communications specialist assigned to HHC/1-27 Infantry Regiment "Wolfhounds." (Photo by CPT Joshua May)

SPC Garrett Simoneau, a native of Winchendon, MA, patrols the Taji Qada, which is northwest of Baghdad, in search of illegal weapons and improvised explosive devices March 14. He assists his unit, Iraqi Security Forces and the Government of Iraq to provide security to the Iraqi people. Simoneau is assigned to A/1-27 Infantry Regiment "Wolfhounds." (Photo by SGT Brad Willeford)

SSG Bruce Hillier, Twin Falls, ID native, and infantryman assigned to B/1-27 Infantry Regiment "Wolfhounds", gives an Iraqi boy candy during a security patrol March 20. The patrols focus on working with the Iraqi Security Forces to continue improving safety in the Taji Qada, northwest of Baghdad. (Photo by SGT Brad Willeford)

SSG Roberto Zuniga, a Laredo, TX native who serves as an infantryman with A/1-27 Infantry Regiment "Wolfhounds", assists an Iraqi boy who cut his finger March 24. (Photo by SPC Brian Pierce)

PFC Daniel Cortes, a Corona, NY native, who serves as an infantryman with Company B, 52nd Infantry Regiment (Anti-Tank) attached to 1-27 Infantry Regiment "Wolfhounds" secures the Hor al Bosh market during a dismounted patrol March 24. Cortes has been assisting the people in the Taji Qada, which is northwest of Baghdad, for nearly four-months by keeping their streets and homes safe from insurgents.
(Photo by SPC Brian Pierce)

SGT Michael Rios, a Sebastian, TX native; SPC Garrett Simoneau, a Winchendon, MA native; and PVT Justin Do, a Vancouver, WA native, enter a school in the Taji Qada, located northwest of Baghdad, March 26. The Soldiers interacted with Iraqi children and distributed needed school supplies. All three Soldiers are assigned to A/1-27 Infantry Regiment "Wolfhounds". (Photo by SGT Brad Willeford)

1LT Louis Cascino, a Cornwall, NY native who serves as a platoon leader with B/1-27 Infantry Regiment "Wolfhounds", conducts physical training with Iraqi Army soldiers at a Joint Security Station in the East Anbar Province, March 28. Coalition Forces are not only training the Iraqi Army on tactical tasks, but physical training, as well.
(Photo by SPC Brian Pierce)

Soldier Inspired by Grandfather's WWII Service

By SPC Brian Pierce
1-27 Infantry Regiment "Wolfhounds"

PVT Edward Robinson, a native of Saint Albans, WV, had his grandfather in mind when he joined the Army in 2007.

Robinson's late grandfather, a World War II veteran, served in the Army as an infantryman. That service is what inspired Robinson to follow in his grandfather's footsteps.

"When you think of the Army, you think of the Infantry," said Robinson.

Robinson is currently assigned to C/1-27 Infantry Regiment "Wolfhounds." He is a member of the Brigade's tactical reserve which rapidly responds to significant activities and events within the brigade's area of operations.

The youngest Soldier in the Battalion, 18-year-old Robinson is helping his unit in ensuring security and stability in the northwest Baghdad area.

In addition to responding to activities on the tactical reserve, he has also provided security for the Brigade's Provincial Reconstruction Team, which assists the Government of Iraq with essential services, such as education for children.

"The Army has helped put so many Iraqi children back in school, and give jobs to the Iraqi people," said Robinson. "The Iraqi people really appreciate what we have done for them, and are starting to improve their lives on their own."

Robinson has quickly learned the Iraqi culture and what it takes to assist in continuing the stabilization of Iraq. His efforts in the area have been rewarded several times by local Iraqi residents. The Iraqi's have shown their gratitude to Robinson by offering him food and smiles.

Robinson said he demands nothing less than perfection from himself. "As soon as he got to the unit, he worked hard to become a member of his team and his unit. He did everything to the best of his ability," said 1LT Ryan Clay, a native of Durham, NC and platoon leader, C/1-27 Infantry Regiment "Wolfhounds."

Robinson said his sense of perfection stems from his grandfather, who passed away when he was twelve. His grandfather's combat experience and stories help Robinson make it through his time in Iraq.

Robinson knows that even though his grandfather is no longer alive, he said he would be proud of all he has accomplished and all he will accomplish.

PVT Edward Robinson, a native of Saint Albans, WV, is assigned C/1-27 Infantry Regiment "Wolfhounds." (Photo by SPC Brian Pierce)

SMITH
NORTON
Cartier

IA, Warriors Search for Caches

By SGT Waldemar Rivera
2nd SBCT PAO, 25th Inf. Div., MND-B

Multi-National Division – Baghdad Soldiers from 3rd Platoon, B/52 attached to 1-27 Infantry Regiment "Wolfhounds" worked alongside Iraqi Army soldiers from the 36th Mechanized Iraqi Army Brigade to search for weapons caches April 17 in Fahad Fadhil, northwest of Baghdad.

The purpose of the mission was to follow-up on reports that indicated the presence of cache sites in the area, to clear the area of possible caches, and show the people of the town the IA is there to protect them.

The mission took place under bad weather conditions caused by an almost blinding sandstorm, which resulted in visibility being reduced to roughly 100 feet all for the combined forces.

"Even with the inclement weather and limited visibility we were facing, we still had to accomplish the mission," said SGT Jason Tippie, a Glouster, OH native, who serves as a squad leader with Company B. "At times, it seemed as if we were walking on Mars."

Despite poor visibility, IA soldiers discovered a bag full of suspicious wires lying in a field near a home.

After a short inquiry on about the origins of the wires, the children pointed to an old air conditioning unit next to the house where the IA soldiers found an unopened box of 7.62 caliber ammunition, a bag full of loose ammunition, higher caliber than normal shot gun ammunition, weapons parts and three AK-47 magazines full of ammo.

Each household is only authorized to have one AK-47 and one magazine unless the occupants have a permit for another weapons system, said SFC Raymond Zwicker, a Concord, CA native, who serves as a platoon sergeant with B/52 Infantry Regimentt.

"When we are in a situation where there are more weapons than allowed, we have to confiscate the excess and hand them over to the IA," he explained.

Inside the house, the IA soldiers found a disassembled AK-47, a jar full of ammunition and six more magazines full of ammo. All excess ammunition and weapons parts were confiscated.

Despite no big finds, the IA soldiers and MND-B Soldiers will continue to work together to clear the area of caches and bring peace of mind to the people in the area.

Iraqi Army soldiers from the 36th Mechanized Brigade show a small munitions cache found in a home April 17 during a joint cache-clearing mission with Soldiers from B/52 attached to 1-27 Infantry Regiment "Wolfhounds," in the town of Fahad Fadhil, northwest of Baghdad. (Photo by SGT Waldemar Rivera)

CPT Brian Sweigart, a native of San Antonio, TX, compares his "guns" with Iraqis in the Taji Qada, northwest of Baghdad April 3. Sweigart is the commander for A/1-27 Infantry Regiment "Wolfhounds." He and his unit continue to work hand-in-hand with local Iraqi leaders to help make Iraq a stronger and more unified nation. (Photo by SGT Brad Willeford)

CPT Kenneth Smith, a native of Phoenix, AZ, inspects an X-ray during a Combined Medical Effort in Nahala Village, in the Taji Qada, northwest of Baghdad April 7. Smith is a physician's assistant assigned to HHC/1-27 Infantry Regiment "Wolfhounds." (Photo by SPC Brian Pierce)

San Antonio native, SSG Daniel Navarro looks for any possible enemy activity while patrolling through the Taji Qada on April 8. Navarro is a reconnaissance scout team leader with HHC/1-27 Infantry Regiment "Wolfhounds." (Photo by CPT Chad Chenoweth)

SPC Richard Bobbitt, a Brenham, TX native, instructs Iraqi Army soldiers on how to enter and clear a hallway during a training exercise April 12 at Camp Taji, northwest of Baghdad. Bobbitt is a mortarman with HHC/1-27 Infantry Regiment "Wolfhounds." (Photo by SPC Brian Pierce)

SPC Allen Post, a San Antonio native, trains Iraqi Army soldiers on basic first aid, during a training event April 12 at Camp Taji, northwest of Baghdad. Post is combat medic for HHC/1-27 Infantry Regiment "Wolfhounds," and has been serving in this region for more than four months helping rid the area of terrorist and criminal activity. (Photo by SPC Brian Pierce)

SSG Vincent Valencia, a Dededo, Guam native, teaches basic medical skills to Iraqi Army soldiers during a training exercise April 12 on Camp Taji, northwest of Baghdad. Valencia servers as a combat medic for HHC/1-27 Infantry Regiment "Wolfhounds."
(Photo by SPC Brian Pierce)

PFC Justin Hollis, a Medina, OH native, trains Iraqi Army soldiers on infantry skills and tactics April 12 during a training event at Camp Taji, northwest of Baghdad. Hollis serves as a mortarman assigned to HHC/1-27 Infantry Regiment "Wolfhounds." (Photo by SPC Brian Pierce)

Erica Corral, of Bensenville, IL, charges a Stryker armored vehicle suspension system with nitrogen on Apr. 15. The Stryker's suspension system is charged with nitrogen to allow for smooth and responsive handling over all types of terrain. Corral is a General Dynamics Land Systems contractor with 1-27 Infantry Regiment "Wolfhounds."
(Photo by CW3 Danny K. Taylor)

A Stryker armored vehicle is lifted on jack stands in preparation for hub testing on Apr. 15. This test was used to detect small cracks in the hub and was part of the semi-annual services conducted on the Strykers. (Photo by CW3 Danny K. Taylor)

SGT Christopher Delaney, of Belleview, FL; and SPC Daniel Stevens, of Dayton, OH, watch carefully as an engine is removed from a Heavy Expanded Mobility Tactical Truck on Apr. 19. Delaney and Stevens are wheeled vehicle mechanics with the 1-27 Infantry Regiment "Wolfhounds." (Photo by CW3 Danny K. Taylor)

SPC Nathaniel Eckhoff, of Sierra Vista, AZ; SPC Donald Ross, of Albuquerque, NM; and SGT Nicolas Laduke, of Minneapolis, MN, perform hub testing on a Stryker armored vehicle on Apr. 20. This test was used to detect small cracks in the hub and was part of the semi-annual services conducted on the Strykers. Eckhoff, Ross, and Laduke are wheeled vehicle mechanics with the 1-27 Infantry Regiment "Wolfhounds." (Photo by CW3 Danny K. Taylor)

SPC Joshua Oden, a San Angelo, TX native, erects an antenna April 12, which enhances his unit's ability to effectively communicate during Operation Gunslinger Shorebreak I, a mission focused at securing the Taji Qada, northwest of Baghdad. Oden is an infantryman assigned to A/1-27 Infantry Regiment "Wolfhounds." (Photo by SGT Brad Willeford)

SGT Michael Rios, a Sebastion, TX native, remains alert, looking for terrorist or criminal activity while patrolling a small village in the Taji Qada, northwest of Baghdad, April 17. Rios serves as an infantry team leader with A/1-27 Infantry Regiment "Wolfhounds." (Photo by SGT Brad Willeford)

With winds picking up and dust blowing around him, SPC Collin Kelly, a Peachtree, GA native, pulls security in the Taji Qada, northwest of Baghdad, April 17. Kelly serves as an infantryman with A/1-27 Infantry Regiment "Wolfhounds," and helps his company and the Iraqi Security Forces secure this area of Iraq. (Photo by SGT Brad Willeford)

CPT Brian Sweigart, a native of San Antonio; LTC Richard "Flip" Wilson, a native of Brooklyn, NY; and CPT Lucas Yoho, a native of Long Beach, CA, prepare to conduct a change of command ceremony April 20. Sweigart relinquished command of A/1-27 Infantry Regiment "Wolfhounds" to Yoho at their Joint Security Station in the Taji Qada, northwest of Baghdad. Wilson is the commander for the 1-27 Infantry Regiment "Wolfhounds." (Photo by SSG JB Jaso III)

CPT Brian Sweigart congratulates 1LT Devon McRainey, of Columbus, OH, for doing an exceptional job leading his Soldiers in combat during a ceremony held at their Joint Security Station in the Taji Qada on April 20. McRainey serves as an infantry platoon leader with A/1-27 Infantry Regiment "Wolfhounds." (Photo by SGT Brad Willeford)

SFC Thomas Ingram, of Mililani, HI, congratulates his Soldiers after affixing their 25th Infantry Division Shoulder Sleeve Insignia-Former Wartime Service, or combat patch for service in a combat zone. Ingram serves as an infantry platoon sergeant with A/1-27 Infantry Regiment "Wolfhounds." (Photo by SGT Brad Willeford)

LTC Richard "Flip" Wilson rewards PFC Samuel Fowler, of Medonough, GA, a commander's coin for excellence during a combat patch ceremony April 20 at a Joint Security Station in the Taji Qada. Fowler is the youngest soldier in A/1-27 Infantry Regiment "Wolfhounds," where he serves as a infantryman. (Photo by SGT Brad Willeford)

CPT Brian Sweigart passes his company's guidon to LTC Richard "Flip" Wilson, officially relinquishing command of A/1-27 Infantry Regiment "Wolfhounds," in the Taji Qada, April 20. Sweigart commanded the "Gunslingers" for more than 21-months and led his men through rigorous training in preparation for their current deployment. Sweigart will assume command of HHC/1-27 Infantry Regiment "Wolfhounds" in July. (Photo by SSG JB Jaso III)

CPT Lucas Yoho, a native of Long Beach, CA, meets a local sheik after assuming command of Multi-National Division-Baghdad's A/1-27 Infantry Regiment "Wolfhounds" April 20. Yoho will work with several local and Iraqi Army leaders to assist them in improving the Taji Qada, northwest of Baghdad. (Photo by SSG JB Jaso III)

SPC Adam Costello, a Northborough, MA native, pulls security during a medical operation in the Taji Qada, northwest of Baghdad, April 21. Costello is a combat medic assigned to HHC/1-27 Infantry Regiment "Wolfhounds." (Photo by SPC Brian Pierce)

SGT Micheal Bird, a Bristol, CT native, directs Iraqis to healthcare providers during a medical operation in the Taji Qada, northwest of Baghdad, April 21. Bird is a combat medic assigned to HHC/1-27 Infantry Regiment "Wolfhounds." (Photo by SPC Brian Pierce)

LTC Richard "Flip" Wilson, commander, 1-27 Infantry Regiment "Wolfhounds," prepares for a flight with Iraqi Army GEN Husan, commander, 36th Iraqi Army Brigade, 9th Iraqi Army Division April 22. The mission was to conduct aerial reconnaissance of their combined area of operations. The two units work together to secure the Taji Qada, northwest of Baghdad. (Photo by SGT Brad Willeford)

A UH-60 Blackhawk helicopter flies over the Taji Qada, northwest of Baghdad, while conducting an aerial reconnaissance of the combined Iraqi Army and Multi-National Division – Baghdad U.S. operational environment April 22. (Photo by SGT Brad Willeford)

SPC Joshua Phillips, a native of Sacramento, CA, uses a metal detector to search for enemy caches in the Taji Qada, northwest of Baghdad, April 25 in temperatures that soared above 105 degrees. Philips serves as an infantryman assigned to A/1-27 Infantry Regiment "Wolfhounds." (Photo by SPC Brian Pierce)

SSG Michael Pharr, Jr., a Charlotte, NC native, observes the impact of the Excalibur round fired during calibration of the M777 Howitzer in the Taji Qada, northwest of Baghdad, April 26. Pharr serves as the fire support non-commissioned officer for A/1-27 Infantry Regiment "Wolfhounds." (Photo by SPC Brian Pierce)

PVT Richard Plasket, of Midlothian, TX, pulls security in the Taji Qada, April 26 in temperatures that soared above 100 degrees. Plasket has been serving in this region northwest of Baghdad as an infantryman with A/1-27 Infantry Regiment "Wolfhounds." (Photo by SSG JB Jaso III)

SSG Michael Pharr, Jr., of Charlotte, NC, plots grid coordinates April 26 to determine the impact point for an Excalibur round fired from the M777 Howitzer. The round was fired from Camp Taji to calibrate the cannon system. Pharr serves as the fire support noncommissioned officer for A/1-27 Infantry Regiment "Wolfhounds." (Photo by SPC Brian Pierce)

1LT Aaron Kletzing, of Chicago, ensures the impact zone is clear April 26 before the firing of an Excalibur round from a M777 Howitzer on Camp Taji, northwest of Baghdad. Kletzing is the fire support officer for A/1-27 Infantry Regiment "Wolfhounds." (Photo by SPC Brian Pierce)

Infantryman's Medic

By SPC Brian Pierce
1st Bn., 27th Inf. Regt., 2nd SBCT, 25th Inf. Div., MND-B

SPC Randel Marcotte, a native of Evergreen, LA, always puts his platoon before himself.

He can't count the number of patrols he's been on, but said " I go on every mission or patrol that involves any guy from my platoon."

No matter how tired he may be, he finds renewed strength to go out every time and be with his guys. Marcotte is a combat medic assigned to B/1-27 Infantry Regiment "Wolfhounds." Marcotte joined the Army, because, "I knew I was needed by my country," he said. "I became a medic so I could come to Iraq to save lives in a time of war, when my country needs me."

Before answering the call, Marcotte grew up as a potato farmer who enjoyed hunting white tail deer in his hometown. In high school he was a member of several clubs helping him to gain good inner personal skills to help people and be involved in things greater than him; giving him a great advantage to interact with the Iraqi people and their leadership in the Taji Qada, northwest of Baghdad.

After high school, Marcotte enrolled as a pre-med student at Northwestern State University of Louisiana; this being the beginning of his interests in the medical field. Being involved in the medical field seems to come natural to him. He even enjoys reading medical books as a hobby to gain medical knowledge.

Marcotte entered the Army on Sept 19, 2005; less than a year later he became a member of the "Wolfhounds," stationed in Hawaii on the Island of Oahu. On Oahu in his free time Marcotte enjoyed Hawaii's great beaches, snorkeling, surfing, and his favorite part, fishing.

Currently deployed northwest of Baghdad, he says he misses the Island's great atmosphere and combination of weather and activities. On his return to the Island he plans to go straight to the North Shore, enjoy the ocean, and then eat sushi at his favorite sushi restaurant, Bonzai Sushi.

When notified about deployment, Marcotte was ready and anxious; he explains that it is what he signed up to do. Deploying would not be easy for Marcotte, knowing he would have to leave his new born son and enter a country he had never been to before.

Currently, Marcotte enjoys helping the people of the Northwest Taji Qada. "Its fun going to small villages that have limited or no contact with coalition forces because it feels good to treat entire villages knowing that they have no other means of medical attention," he said.

Marcotte's platoon sergeant, SFC Stephen Novak, a McAllen, TX, native, says, "Marcotte is a hardworking Soldier, who takes care of the guys in the platoon."

SGT Malaefou Heather, a Leone, American Samoa native continues to praise Marcotte's efforts by saying that he "is willing to learn new things and train with the Soldiers and other medics." These are strong qualities that help Marcotte be a good medic.

His platoon has found a high numbers of enemy weapon caches. They have taken contact from an improvised explosive device during a patrol in their area of operations, with no coalition forces injured or killed. They have continued to rid the area of terrorist activity, and have been assisting the Government of Iraq in providing much-needed essential services, such as medical aid, potable water, electricity, and education.

While Marcotte continues his duties and undying commitment as a Wolfhound combat medic, he never stops thinking about his wife and one-year-old son back home. Marcotte's commitment to the Global War on Terrorism is an example that his son will look-up to several years from now.

On Apr. 22, Evergreen, LA native, SPC Randel Marcotte treats an Iraqi boy during a medical operation, focused at assisting residents in the Taji Qada that have no access to medical attention. Marcotte is a combat medic assigned to B/1-27 Infantry Regiment "Wolfhounds." (Photo by SPC Brian Pierce)

SPC Joshua Phillips, a Sacramento, CA native, takes cover from the sun in temperatures soaring over 100 degrees, while on patrol in the Taji Qada, northwest of Baghdad, April 26. Phillips is an infantryman assigned to A/1-27 Infantry Regiment "Wolfhounds." (Photo by SSG JB Jaso III)

CPT Shaun Wheelwright, of Pinetop, AZ, carries a box of school supplies donated by Hawaii-based company, "Oils of Aloha," into a school in the Taji Qada, April 26. Wheelwright is the assistant operations officer for the 1-27 Infantry Regiment "Wolfhounds." (Photo by SSG JB Jaso III)

CPT Shaun Wheelwright, of Pinetop, AZ, waves hello in a sea of children at a school in the Taji Qada, April 26. The children just received school supplies donated by Hawaii-based company, "Oils of Aloha." Wheelwright is the assistant operations officer for the 1-27 Infantry Regiment "Wolfhounds." (Photo by SSG JB Jaso III)

SPC Joshua Phillips, of Sacramento, CA, carries a box of school supplies, donated by Hawaii-based "Oils of Aloha," to a school April 26 in a remote village in the Taji Qada. Phillips is an infantryman assigned to A/1-27 Infantry Regiment "Wolfhounds" and has been serving in Iraq for five months.
(Photo by SSG JB Jaso III)

SPC Kristopher Weber, a Rosehill, KS native, patrols an area of the Taji Qada, northwest of Baghdad, April 28. Weber is a combat medic assigned to B/1-27 Infantry Regiment "Wolfhounds." (Photo by SSG JB Jaso III)

PFC Timothy Brown, a native of Cedar Rapids, IA, patrols a market in the Taji Qada, northwest of Baghdad, April 28, ensuring growth of this economic district remains free of criminal and terrorist activity. Brown is an infantryman assigned to B/1-27 Infantry Regiment "Wolfhounds." (Photo by SSG JB Jaso III)

SPC Wayne Cloud, Jr., a Sisseton, SD native, assists the Iraqi Army in securing a marketplace in the Taji Qada, northwest of Baghdad, April 28. Cloud is a forward observer assigned to B/1-27 Infantry Regiment "Wolfhounds." (Photo by SSG JB Jaso III)

SPC Wesley Shipman, a native of Katy, TX, secures a business district in the Taji Qada, northwest of Baghdad, April 29, allowing shoppers to purchase needed items safely without fear of criminal or terrorist activities. Shipman is an infantryman assigned to B/1-27 Infantry Regiment "Wolfhounds." (Photo by SSG JB Jaso III)

SFC James Burciaga, of West Covina, CA, balances on a log while crossing a field in the Al Awad region, northwest of Baghdad, on May 9. Burciaga is an infantryman platoon sergeant with A/1-27 Infantry Regiment "Wolfhounds." (Photo by PFC Stephen Miller)

SPC George Davis, of Buena Vista, CO, is joined with an Iraqi Army soldier in the Abu Ghaib Qada on May 2. Davis is an infantryman with B/52 Infantry Regiment "Hammer," attached to the 1-27 Infantry Regiment "Wolfhounds." (Photo by CPT John Goldsworthy)

CPT David Uthlaut, of Charleston, SC; CPT John Goldsworthy, of Rosalia, WA; and MAJ Glen Helberg at a Sheik Meeting in the Abu Ghaib Qada on May 2. Uthlaut commands C/1-27 Infantry Regiment "Wolfhounds;" Goldsworthy commands B/52 Infantry Regiment "Hammer;" and Helberg commands C/1-21 Infantry Regiment "Gimlets." (Courtesy Photo)

SSG Curtis Johnson, of Waterport, NY, makes sure the people have a safe community in which to conduct business in the Taji Qada, May 3. Johnson serves as an infantry squad leader with A/1-27 Infantry Regiment "Wolfhounds." (Photo by SGT Brad Willeford)

SSG Daniel Ramirez, a Moreno, CA native; and PFC Jesse Weber, a Richland Center, WI native, enjoy a bit of shade while pulling security on a patrol to check out several of the local businesses in the Taji Qada, located northwest of Baghdad, May 3. Ramirez and Weber serve as infantrymen with A/1-27 Infantry Regiment "Wolfhounds." (Photo by SGT Brad Willeford)

PFC Scott Blazek, a Pewaukee, WI native, watches for any enemy activity while conducting a dismounted patrol in the Taji Qada, May 6. Blazek is a infantryman assigned to A/1-27 Infantry Regiment "Wolfhounds." (Photo by SSG JB Jaso III)

PFC Stephen Miller, of Batesville, AR, prepares for a dismounted patrol, which is focused on interacting with residents of a small remote village in the northern area of Taji Qada, May 6. Miller is an infantryman assigned to A/1-27 Infantry Regiment "Wolfhounds." (Photo by SSG JB Jaso III)

SPC Arthur Smith, of Sasakwa, OK, pulls security in the northern area of the Taji Qada during a dismounted patrol on May 6. Smith is an infantryman assigned to A/1-27 Infantry Regiment "Wolfhounds." (Photo by SSG JB Jaso III)

SGT Douglas Urias, a Houston native, pulls security so his fellow Soldiers can check on local businesses in the Taji Qada, located northwest of Baghdad, May 3. Urias serves as an infantry team leader with A/1-27 Infantry Regiment "Wolfhounds." (Photo by SGT Brad Willeford)

Wolfhound Makes Difference Training Iraqi Army Soldiers

By SPC John Ahn
1-27 Infantry Regiment "Wolfhounds"

Working together with the Iraqi Army is an important step in the success of this war – and an important step in Iraqis taking charge of their nation.

A fire team showing a unified coalition front line force at a Joint Security Station near Camp Taji, northwest of Baghdad, is a model for a successful transition from a dependent Iraq to a sovereign nation that can secure the people.

The fire team consists of both Multi-National Division – Baghdad Soldiers and IA soldiers from the 36th Brigade, 9th Iraqi Army Division. SGT Michael Rios, a native of Sebastian, Texas, serves as the team chief. He is assigned to 2nd Platoon, A/1-27 Infantry Regiment "Wolfhounds."

The formation of the combined group initially came together out of sheer necessity when Rios' Soldiers were reassigned to other duties or left the area for Environmental Morale Leave.

The sudden moves forced the U.S. and Iraqi Soldiers to learn to work together and overcome linguistic barriers. Overcoming those barriers was a challenge, said Rios, but they were eased through the use of interpreters and hand signals.

"The most amazing aspect of the fire team is that we speak so little to each other," Rios said. "We work mostly by hand signals and knowing glances."

Acknowledgement, he added, is made by a simple nod.

"They work like any other fire team in a modern army," he said. "They know their jobs, take directions from their leader and execute. They are well rehearsed and work flawlessly together."

Rios said he believes in their ability to work as members of a team. In fact, he added, they train together every chance they get. During their available free time, they come together to work on close-quarters combat drills, practice search techniques for vehicles and buildings, and constantly refine their battle-drill skills. All team members know what they are expected to do when hostile contact is made.

"He is a good leader," stated Hamed, an IA Soldier from the 36th Bde. Hamed has been on Rios' team since the beginning. "(He's) patient and decisive – and very funny too."

Even with a language barrier, Rios is passing on his knowledge and preparing the IA soldiers to be future leaders. He said his vision is to be a force multiplier by being an example so that these soldiers can one day show their future soldiers how to be proficient warriors.

Rios said he demands excellence from his team and doesn't allow them to give less than their best.

"I want 100 percent from them, their undivided attention. If they work hard now and train properly, when it comes to firefights and real-life matters, they won't have to think about it," he said. I want them to identify what they must to do and execute."

The tough training and expectations have brought the Soldiers of these two diverse cultures together to form a close-knit team and have become brothers in arms.

SGT Michael Rios, a Sebastian, TX native, pulls security next to an Iraqi Army soldier in his team in the Taji Qada, northwest of Baghdad, May 6. Rios, not only patrols with Iraqi Army Soldiers, but continues to train with them. He is an infantry team leader assigned to A/1-27 Infantry Regiment "Wolfhounds." (Photo by SSG JB Jaso III)

SGT Michael Rios, a native of Sebastian, TX, patrols a northern remote village in the Taji Qada, located northwest of Baghdad, May 6. Rios is a team leader assigned to A/1-27 Infantry Regiment "Wolfhounds." His team is composed of MND-B Soldiers and Iraqi Army soldiers from the 36th Brigade, 9th Iraqi Army Division. (Photo by SSG JB Jaso III)

SGT Michael Rios, a native of Sebastian, TX, leads his team through a house in the Taji Qada, where they conducted searches to locate enemy weapons and activity May 6. Rios is an infantry team leader assigned to A/1-27 Infantry Regiment "Wolfhounds." His team is composed of MND-B Soldiers and Iraqi Army soldiers from the 36th Brigade, 9th Iraqi Army Division. (Photo by SSG JB Jaso III)

SPC Joseph Baker, a Blanco, TX native, walks the streets of Iraq to clear them of improvised-explosive devices so others may travel safely during Operation Gunslinger Shorebreak II in the Taji Qada, located northwest of Baghdad, May 8. Baker serves as a forward observer with A/1-27 Infantry Regiment "Wolfhounds." (Photo by SGT Brad Willeford)

LTC Richard "Flip" Wilson, a Brooklyn, NY native, discusses safety and essential services with local sheiks and community leaders during a meeting in the East Anbar Province, northwest of Baghdad, May 8. Wilson is the commander for 1-27 Infantry Regiment "Wolfhounds." (Photo by SSG JB Jaso III)

PVT Christian Diaz, a Miami native, takes time to talk with children while pulling security at a meeting between military and local leaders in the East Anbar region, northwest of Baghdad, May 8. Diaz serves as an infantryman with HHC/1-27 Infantry Regiment "Wolfhounds." (Photo by SSG JB Jaso III)

SPC Timothy Russell, of Yardley, PA; 1LT Ryan Clay, of Durham, NC; and SPC Eric Waddle, of Bedford, IN, enjoy Iraqi cuisine provided by a local sheik during a meeting where military and local leadership discussed security of the East Anbar province on May 8. Russell and Waddle are mortarmen assigned to HHC/1-27 Infantry Regiment "Wolfhounds." Clay is a mortar platoon leader assigned to HHC/1-27 Infantry Regiment. (Photo by SSG JB Jaso III)

1LT Ryan Clay, a native of Durham, NC, does magic tricks and jokes with local residents while visiting a sheik's home in the East Anbar province, northwest of Baghdad, May 8. Clay is a mortar platoon leader assigned to HHC/1-27 Infantry Regiment "Wolfhounds."
(Photo by SSG JB Jaso III)

1LT Nicholas Pines, a Pinellas Park, FL native, rapidly receives reports from units during a significant activity in his unit's operational environment May 9. Pines is the task force battle captain assigned to HHC/1-27 Infantry Regiment "Wolfhounds," which is headquartered at Camp Taji, northwest of Baghdad. (Photo by SSG JB Jaso III)

CPT Arnulfo Martinez, a San Antonio native, communicates with leaders at a Joint Security Station in the Taji Qada, northwest of Baghdad, regarding significant activities in the unit's area of operations May 9. Martinez is the commander for HHC/1-27 Infantry Regiment "Wolfhounds."
(Photo by SSG JB Jaso III)

CPT David Uthlaut, a Charleston, SC, native, and commander of C/1-27 Infantry Regiment "Wolfhounds," addresses his Soldiers May 6 at Camp Taji. The Soldiers recently returned from 42 days in the Sadr City district of Baghdad, where they reinforced Iraqi Army soldiers in stabilizing the city. Uthlaut praised his Soldiers for their dedication and hard work during combined operations with the Iraqi Army. (Photo by SSG JB Jaso III)

CPL Matthew Dortch, of Redding, CA, is awarded a Combat Medical Badge by LTC Richard Wilson, commander, 1-27 Infantry Regiment "Wolfhounds," during a ceremony at Camp Taji on May 6. Dortch, a combat medic assigned C/1-27 Infantry Regiment "Wolfhounds," was presented the CMB for providing medical aid to Soldiers during fierce fighting in Sadr City in April. (Photo by SSG JB Jaso III)

PVT Edward Robinson, of Saint Albans, WV, is awarded a Combat Infantryman Badge by COL Todd McCaffrey, commander, 2nd Stryker Brigade Combat Team, 25th Infantry Division, at Camp Taji on May 6. Robinson, an infantryman assigned to C/1-27 Infantry Regiment "Wolfhounds," was awarded the CIB for serving as an infantryman during fierce fighting in the Sadr City district of Baghdad for more than 40 days. (Photo by SSG JB Jaso III)

SGT Steven Walters, of Wichita, KS, is awarded the Purple Heart May 6 by COL Todd McCaffrey, commander, 2nd Stryker Brigade Combat Team, 25th Infantry Division, at Camp Taji for injuries sustained when his Stryker armored vehicle was hit by an improvised explosive device Jan. 18. Despite his injuries, Walters continued to serve as an infantry team leader with C/1-27 Infantry Regiment "Wolfhounds." (Photo by SSG JB Jaso III)

CPL Kristopher Weber, a Rosehill, KS native, takes time to listen to what the locals both young and old have to say in the Taji Qada, northwest of Baghdad, May 9. Weber serves as a medic with B/1-27 Infantry Regiment "Wolfhounds." (Photo by SGT Brad Willeford)

SGT Jeffrey Coots, of Brockton, MA, talks with Iraqi Army soldiers at a checkpoint in the Taji Qada, May 9. He discussed security and medical capabilities in the region. Coots is a combat medic assigned to HHC/1-27 Infantry Regiment "Wolfhounds." (Photo by SSG JB Jaso III)

1LT Ryan Clay, of Durham, NC, flashes the "shaka" with Iraqi Army soldiers at a checkpoint in the Taji Qada May 9. Clay discussed security measures in this region. He is a mortar platoon leader assigned to HHC/1-27 Infantry Regiment "Wolfhounds." (Photo by SSG JB Jaso III)

SPC Aarron Morris, of Leesville, SC, enjoys a laugh while on patrol in the Taji Qada, May 10. Morris serves as an infantryman with B/52 Infantry Regiment, attached to 1-27 Infantry Regiment "Wolfhounds." (Photo by SGT Brad Willeford)

CPL Christopher Tinkham, of Tonawanda, NY, plays soccer with children during a patrol in the East Anbar Province, northwest of Baghdad, May 11. Tinkham is a mortarman assigned to HHC/1-27 Infantry Regiment "Wolfhounds." (Photo by SSG JB Jaso III)

SSG Chad Ryan, of Sterling, IL, is greeted by a cow while taking a break during a patrol May 12 in the East Anbar province, northwest of Baghdad. Ryan is a mortar section sergeant assigned to HHC/1-27 Infantry Regiment "Wolfhounds." (Photo by SSG JB Jaso III)

The sun shines into the eyes of SPC Jacob Bruno, of Randolph, MN, as he assists in securing a building where possible terrorist activities have taken place during Operation Raider Riptide I, in the Tarmiya Qada, May 15. The operation focused on capturing high-value individuals that have caused terror to the residents in this region, located northwest of Baghdad. Bruno is a combat medic assigned to HHC/1-27 Infantry Regiment "Wolfhounds."
(Photo by SSG JB Jaso III)

SFC James Burciaga, of West Covina, CA, observes SPC Joshua Lloyd, of Rocky Ridge, MD, fire a .50cal machine gun atop of a Stryker armored vehicle, during a qualification range on May 17. Both soldiers are infantrymen with A/1-27 Infantry Regiment "Wolfhounds." (Photo by PFC Stephen Miller)

LTC Richard "Flip" Wilson listens to security concerns from leaders from the 36th Brigade, 9th Iraqi Army Division. The IA and Soldiers from 1-27 Infantry Regiment "Wolfhounds," work together to ensure the safety and security of the Taji Qada, northwest of Baghdad.
(Photo by SSG JB Jaso III)

CPT Omololu Makinde, a Columbus, OH native, discusses locations of possible enemy activity in the Taji Qada, northwest of Baghdad, with leaders from the 2nd Battalion, 31st Brigade, 8th Iraqi Army Division, May 19. Makinde is the Iraqi Security Forces coordinator for the 1-27 Infantry Regiment "Wolfhounds."
(Photo by SSG JB Jaso III)

SPC Kenneth Severin, a Mesa, AZ native, listens to instructions during a patrol focused at disrupting terrorist and criminal movement in the Taji Qada, northwest of Baghdad, May 20. Severin is a combat medic assigned to HHC/1-27 Infantry Regiment "Wolfhounds."
(Photo by SSG JB Jaso III)

PFC Joseph Kendall, a native of Wayland, NY, secures a vehicle checkpoint, while the Soldiers in his unit conduct vehicle inspections deterring movement of enemy weapons in the Taji Qada, located northwest of Baghdad, May 20. Kendall is an infantryman assigned to A/1-27 Infantry Regiment "Wolfhounds."
(Photo by SSG JB Jaso III)

SSG Roberto Zuniga, a Laredo, TX native, greets a child while on patrol in the Taji Qada, northwest of Baghdad, May 10. He and his unit continue to build positive relationships with the people in the area. Zuniga serves as an infantry squad leader with A/1-27 Infantry Regiment "Wolfhounds." (Photo by SGT Brad Willeford)

A child curiously approaches SPC Riki Tate, a San Francisco native, when he takes a break from walking the streets of the Taji Qada, northwest of Baghdad, May 13. Tate said he enjoys taking time during his patrol to stop and talk with children in the area. Tate has been working in the area for the last five months as an infantryman with B/52 Infantry Regiment "Hammer." (Photo by SGT Brad Willeford)

SFC Raymond Zwicker, a Modesto, CA native, leads his men on a patrol through the streets of Iraq while making assessments of the local security check points in the Taji Qada, northwest of Baghdad, May 13. Zwicker serves as a platoon sergeant with B/52 Infantry Regiment, attached to 1-27 Infantry Regiment "Wolfhounds." (Photo by SGT Brad Willeford)

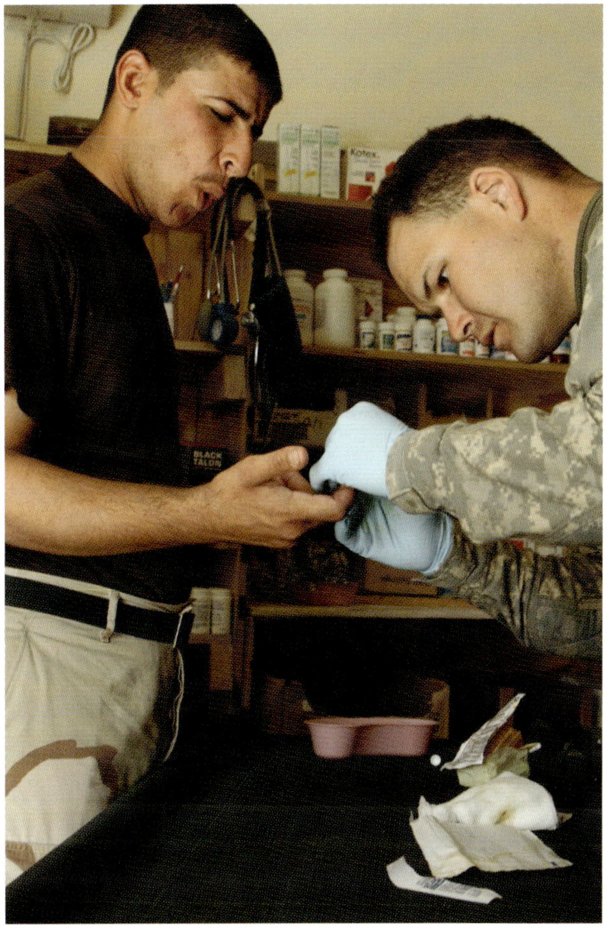

SGT Mark Davenport, a Wimberly, TX native, treats an injury on an Iraqi Army Soldier's hand at Camp Taji, northwest of Baghdad, May 16. Davenport, the senior medic for A/1-27 Infantry Regiment "Wolfhounds," uses his medical skills to improve the lives of the people of Iraq. (Photo by SGT Brad Willeford)

CPL Christopher Tinkham, a Tonawanda, NY native; and SGT Robert Lacy, a Bandera, TX native, work with the Taji residents to provide security for the Northwest Taji Sheik Support Council meeting in the Taji Qada, located northwest of Baghdad, May 17. Tinkham and Lacy serve as mortarmen with HHC/1-27 Infantry Regiment "Wolfhounds." (Photo by SGT Brad Willeford)

With a child on his shoulders, 2LT Joshua Kordela, of Owings Mills, MD, waits with neighborhood children outside a medical operation in Hatim Mutliq, northwest of Baghdad, May 22. Kordela said he enjoys working with the people of Iraq and doing his part to help build a better Iraq. Kordela serves as an infantry platoon leader with B/52 Infantry Regiment "Hammer." (Photo by SGT Brad Willeford)

Wolfhound multi-tasks to keep Iraq safe

By CPT Brian Sweigart
1st Bn., 27th Inf. Reg., 2nd SBCT, 25th Inf. Div., MND-B

Since September 11, 2001, New Yorkers have been at the epicenter of the war on terror, and a Multi-National Division Baghdad Soldier is continuing that tradition.

The Soldier is SPC Michael Ortiz, a native of Riverhead, NY, who deployed in support of Operation Iraqi Freedom and is conducting missions daily to secure the citizens of Iraq; he is an infantryman assigned to A/1-27 Infantry Regiment "Wolfhounds."

Close to 2.5 years ago, Ortiz enlisted in the Army to serve as an infantryman. He said his father and his step-father's service as a Seabee and infantrymen influenced him to join the military. Upon arriving at Schofield Barracks, Hawaii, in Dec. 2005, he was assigned to A Co. and immediately began training and taking in the island life.

"Before deploying, I was really into surfing and hanging out with the guys on the beach," Ortiz stated, about his time on Hawaii. "When I get back, I want to get better at surfing and take my Denny Pang board out to Sunset Beach."

When it came to training for the deployment, he said he took no short cuts and strived to be the best. His motivation and desire to learn resulted in his selection to serve as one of the company marksmen prior to the deployment.

"In May 2007, I went to … school at Fort Benning, (GA), and it was one of the best schools I have been to in the military," Ortiz said.

His hard work and training finally culminated in Dec. when his unit deployed to the Taji Qada, northwest Baghdad.

"I expected it, so I was ready," he said. "I was not nervous, and I was excited to see a new culture."

Thus far, the deployment has kept Ortiz busy; in addition to being employed in his assigned role, Ortiz has also worked on the personal security detachment for the commander and as a unmanned aerial vehicle operator for the company.

"On (our) missions, we over-watch areas of interest for the company and look for improvised-explosive device emplacement and suspicious activity," said Ortiz, adding that he typically operates at night.

Beyond the over-watch missions, he has served as an integral part of other security missions in the Taji area, to include the capture of two high-value individuals and multiple weapons caches, consisting of rocket-propelled grenades and mortars.

Oritz is "very dedicated to the mission and always accomplishes it with minimal supervision," said CPT Lucas Yoho, a native of Long Beach, CA, who serves as the commander of A Co., in talking about the multiple tasks that Ortiz is given. "He's a team player."

The multitudes of missions have also brought SPC Ortiz into close contact with local residents in the area.

"The locals have all been friendly, but the language barrier gets in the way a lot," he said.

One area that the language has not impeded the cross-cultural experience for Ortiz is with the food. "I've tried the Iraqi cuisine, as they always offer it to us first as their guests," he said. "It's good, but it's not like New York Pizza!"

While deployed, Ortiz is continuing to look toward the future by taking college classes online and preparing for his next move. He is considering staying in the military with a goal of becoming a special forces operator or warrant officer.

In the meantime, he said he continues to carry on the tradition of excellence of New Yorkers and patrols the streets to help secure the people of Iraq. At the end of the road, he knows that Riverhead and his Family will be waiting and ready for his safe return.

SPC Michael Ortiz, a Riverhead, NY native assists the Iraqi Army May 18 in securing the northern region of the Taji Qada, Ortiz is an infantryman assigned to A/1-27 Infantry Regiment "Wolfhounds." (Photo by SSG JB Jaso III)

Along with several tasks involved with helping to keep the Taji Qada free of terrorist activity, SPC Michael Ortiz, a native of Riverhead, NY, continues to excel at taking college courses while deployed to Camp Taji, located northwest of Baghdad. Ortiz is an infantryman assigned to A/1-27 Infantry Regiment "Wolfhounds." (Photo by SPC Brian Pierce)

SFC Ernest Oberst, a native of Gladstone, MI, discusses plans to deter enemy freedom of movement with an Iraqi Army soldier, in the Taji Qada, northwest of Baghdad, May 20. Oberst is a platoon sergeant assigned to A/1-27 Infantry Regiment "Wolfhounds."
(Photo by SSG JB Jaso III)

SGT David Exton, a native of Pocatello, OH, patrols the northern region of the Taji Qada May 20, to ensure the safety of the residents in the region, located northwest of Baghdad. Exton is an infantry team leader assigned to A/1-27 Infantry Regiment "Wolfhounds."
(Photo by SSG JB Jaso III)

SSG Joseph Simpliciano, a Kapolei, HI native, makes a sweep of a field in search of weapons caches during Operation Gunslinger Bonzai III in the Taji Qada, located northwest of Baghdad, May 20. Simpliciano serves as a mortar squad leader with A/1-27 Infantry Regiment "Wolfhounds." (Photo by SGT Brad Willeford)

SPC Charles Joseph Jr., of Kerrville, TX, patrols the northern region of the Taji Qada on May 20 to ensure that the residents in this area remain free of terrorist and criminal activities to provide the Government of Iraq the opportunity to safely provide essential services. Joseph is a radio telephone operator assigned to A/1-27 Infantry Regiment "Wolfhounds."
(Photo by SSG JB Jaso III)

SGT Eberardo Solis, a Mission, TX native, carries a can of ammunition found buried during Operation Gunslinger Bonzai IX, in the Taji Qada, northwest of Baghdad, May 27. The mission aimed to clear the area of enemy weapon caches. Solis serves as an infantry team leader with A/1-27 Infantry Regiment "Wolfhounds." (Photo by SGT Brad Willeford)

As the sun sets, MAJ Michael Zernickow, a Concord, CA native, prepares to move toward a house that is suspected to have terrorists inside during an operation focused at removing terrorist activity from the Taji Qada, northwest of Baghdad, May 22. Zernickow is the operations officer for the 1-27 Infantry Regiment "Wolfhounds." (Photo by SSG JB Jaso III)

SGT Timothy Hobbs, a Wilmington, NC native, escorts a suspected terrorist during an operation focused at removing terrorist activity from the Taji Qada, northwest of Baghdad, May 22. Hobbs is a sniper team leader, assigned to HHC/1-27 Infantry Regiment "Wolfhounds." (Photo by SSG JB Jaso III)

CPT Chad Chenoweth, a native of Hamilton, OH, enjoys pastries and chai while discussing safety and security with a Sheik in the Taji Qada, northwest of Baghdad, May 25. Chenoweth is the reconnaissance platoon leader assigned to HHC/1-27 Infantry Regiment "Wolfhounds." (Photo by SSG JB Jaso III)

SPC Garrett Simoneau, a Wichendon, MA native, walks past residents on his way to clear a house in the Taji Qada, northwest of Baghdad, May 27. Simoneau serves as an infantryman with A/1-27 Infantry Regiment "Wolfhounds." (Photo by SGT Brad Willeford)

SPC Adam Bolderson a native of Kansas City, MO, helps to secure the Taji Qada, northwest of Baghdad, May 28 to ensure the region remains free of terrorist and criminal activity. Bolderson is a mortarman assigned to HHC/1-27 Infantry Regiment "Wolfhounds." (Photo by SSG JB Jaso III)

SGT Malaefou Heather, a native of Leone, American Samoa, patrols a market area in the Taji Qada May 28, as shoppers and entrepreneurs safely conduct business without worrying about terrorist or criminal activity. Heather is a combat medic assigned to HHC/1-27 Infantry Regiment "Wolfhounds." (Photo by SSG JB Jaso III)

SPC Austin Sandoval, a native of Lukachukai, AZ, patrols the perimeter of a checkpoint to help ensure enemy forces do not infiltrate this position in the Taji Qada May 28. Sandoval is a mortarman assigned to HHC/1-27 Infantry Regiment "Wolfhounds." (Photo by SSG JB Jaso III)

SPC Collin Kelly, a Peachtree City, GA native, oversees the payday activities for the Sons of Iraq (Abna al-Iraq) guards in the Taji Qada, northwest of Baghdad, May 28. Kelly serves as an infantryman with A/1-27 Infantry Regiment "Wolfhounds." (Photo by SGT Brad Willeford)

SMITH
DAVENPORT
DAVENPORT

Community heals through education

By PFC John Ahn
1st Bn., 27th Inf. Reg., 2nd SBCT, 25th Inf. Div., MND-B

It was a picturesque day in the Taji Qada, northwest of Baghdad, perfect for the opening of al Thoha school, June 5.

There was a jovial feeling in the air and it was evident in the faces of the young and old alike on the clear sky afternoon.

The headmaster of al Thoha School, Akmed Salmin Hussein, stated, "Al Qaeda destroyed this school [two years ago], but with the help of coalition forces this school was rebuilt. This area has been made safe by the Iraqi Army and the presence of coalition forces."

Soldiers from 1-27 Infantry Regiment "Wolfhounds," work to secure this area of the Taji Qada and have had a presence for over six months. Their presence, along with members of the Iraqi Security Forces was instrumental in the rebuilding and opening of the school.

This community has begun to heal from the war that has consumed this nation.

"We're at the tipping point," said 1LT Erik Wiesenhan, a native of Camby, OR, and member of the embedded Local Reconstruction Team for the Wolfhounds. "It's really a unique time now. The locals can take pride that it's being funded by the Government of Iraq and ultimately by them."

Iraq's Ministry of Education has begun to prompt change for the children and the people of Iraq as a whole.

As the people of Iraq are beginning to take a more active role in dictating their future, the Government of Iraq and the ISF are there to shoulder the weight as a foundation that is well-built.

"It went well. The people who needed to be there were there," stated 2LT John Busuego, a native of Ontario, CA, and education advisor for the Wolfhounds eLRT. "The GoI, Iraqi Army, local representatives were all involved. There's really a sense of accomplishment."

This community has begun to heal, and the Wolfhounds continue to secure and assist the Government of Iraq in reestablishing essential services, such as education, to the residents in the area.

Local leaders and Government of Iraq officials took part in the ribbon cutting ceremony, marking the official re-opening of the al Thoha School in the Taji Qada June 5. The school was destroyed by terrorists two years ago. Security was provided by the 2nd Battalion, 31st Brigade, 8th Iraqi Army Division and 1-27 Infantry Regiment "Wolfhounds." (Photo by SSG JB Jaso III)

SGT Robert Lacy, a Bandera, TX native, patrols a marketplace in the East Anbar Province June 4, while leaders discuss future market projects that would assist in improving the economy of this region. Lacy is a mortar squad leader assigned to HHC/1-27 Infantry Regiment "Wolfhounds." (Photo by SSG JB Jaso III)

SFC James Burciaga, of West Covina, CA, is joined with children in the Al Awad region on Jun. 5. Burciaga is an infantry platoon sergeant assigned to A/1-27 Infantry Regiment "Wolfhounds." (Photo by PFC Stephen Miller)

SPC Alexignacio Trinidad, a Solvang, CA native, talks with residents in the East Anbar Province informing them of how to better secure their neighborhood, June 6. Trinidad serves as an infantryman with B/52 Infantry Regiment "Hammer." (Photo by SGT Brad Willeford)

1LT Devon McRainey, a native of Cleveland, OH, takes time to listen to an elderly man during a weapons cache search in the Taji Qada, northwest of Baghdad, June 2. McRainey is an infantry platoon leader assigned to A/1-27 Infantry Regiment "Wolfhounds." (Photo by SGT Brad Willeford)

Wolfhound is a renaissance Soldier

By PFC John Ahn
1st Bn., 27th Inf. Regt., 2nd SBCT, 25th Inf. Div., MND-B

Being a Soldier means being a master at many things and having the ability to adapt to any environment. For one Soldier from HHC/1-27 Infantry Regiment "Wolfhounds," adapting to his environment has been a way of life.

Such is the case for SSG Hiram Barbosa, a native of Mayaguez, Puerto Rico, who is a Stryker armored vehicle commander and is responsible for the vehicle crews for the Wolfhound commander's Personal Security Detail.

Barbosa, along with his seven brothers and sisters grew up very poor in Puerto Rico. Their mother was single and worked at a local bar. His life changed dramatically when his mother died when he was 16. It was now time for him to adapt to his new environment as he moved to Berkeley Heights, NJ, to live with his godparents.

Barbosa said his godparents had an immense impact on how he lives his life today. He is very focused on his health and fitness, which is something he learned from them. His godfather is a professional body builder and a police officer, and his godmother is an aerobics and fitness instructor.

Their active lifestyles were such an influence that he recently competed in a body building contest and earned third place at Camp Taji, which is located northwest of Baghdad, May 25.

The balance he keeps in fitness, along with his daily mission tempo, is nothing short of amazing. He said speaking to his godfather helps him keep that balance. In fact, he calls him as often as he can.

"He's a great influence in my life," said Barbosa, adding that his godfather constantly wants to help him improve himself in every aspect of his life.

"He was a happy kid," said Tony Martinez, Barbosa's godfather. "He was the type of kid, you tell him what to do, and he did it. He was never afraid, always positive, and he never gave up."

His dedication for fitness carries over to his Soldiers. His driver, SPC Daniel Van Houten, a native of Pasadena, TX, recently began working out with Barbosa and said he has seen vast improvements in both his physique and endurance.

"He pushes me," said Van Houten, regarding Barbosa's influence to improve his physical fitness. "He trains me properly, and he motivates not only me but the whole section."

Barbosa's dedication to his Family is what drove him to the Army life. He enlisted immediately after graduating from Jose de Diego High School in New Jersey.

He said he chose to enlist in the Army as an infantryman to fulfill his yearning to "blow things up, shoot weapons and advance quickly" in his career field.

"I was a little bit fearful, but he's a tough kid," said Martinez, who said he was initially apprehensive about Barbosa joining the military. "I knew he could handle it."

He is currently serving in his fourth deployment since joining the Army, which includes two tours in Iraq and a tour in both Afghanistan and Bosnia.

He said he plans to make the Army his career and continues to become more tactically and technically proficient on military tasks, which improves his chances to be promoted through the ranks.

SSG Hiram Barbosa, a native of Mayaguez, Puerto Rico, flexes during the final pose of the Camp Taji Fitness Competition at Camp Taji May 25. Barbosa placed third and attributes his interest in fitness to his godparents. Barbosa is a Stryker armored vehicle commander assigned to HHC/1-27 Infantry Regiment "Wolfhounds." (Photo by PFC John Ahn)

SSG Hiram Barbosa, a native of Mayaguez, Puerto Rico, loads a .50-caliber heavy machine gun prior to departing Camp Taji on a patrol into the Taji Qada June 3. Barbosa is a Stryker armored vehicle commander assigned to HHC/1-27 Infantry Regiment "Wolfhounds."
(Photo by PFC John Ahn)

SSG Hiram Barbosa, a native of Mayaguez, Puerto Rico, conducts a single back bicep pose during the Camp Taji Fitness Competition at Camp Taji May 25. Barbosa is a Stryker armored vehicle commander assigned to HHC/1-27 Infantry Regiment "Wolfhounds."
(Photo by PFC John Ahn)

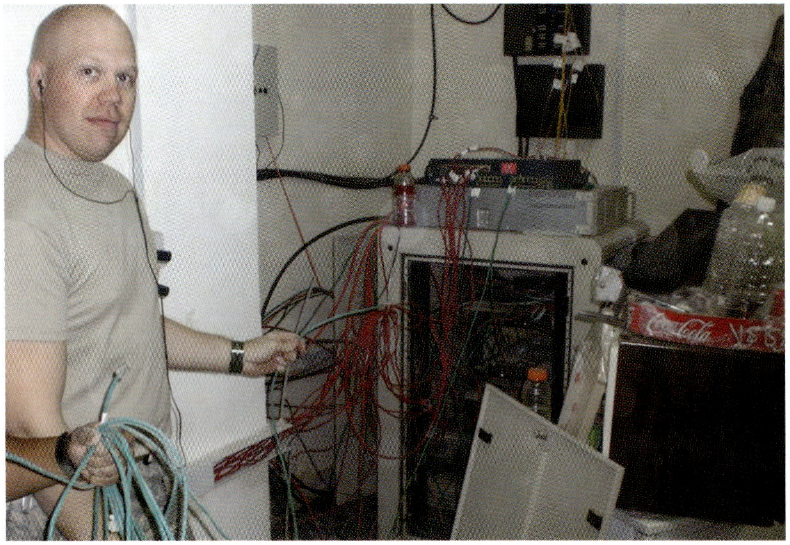

Cuyahoga, OH native, SPC Aaron Markwalder organizes wiring for servers at the battalion's communications office at Camp Taji on June 6. Markwalder is a network systems specialist assigned to HHC/1-27 Infantry Regiment "Wolfhounds." (Photo by CPT Joshua May)

Brockton, MA native, SGT Jeffrey Coots cuts a sheet of plywood in order to build cabinets and tables for the communications equipment at Camp Taji on June 6. Coots is a combat medic assigned to HHC/1-27 Infantry Regiment "Wolfhounds." (Photo by CPT Joshua May)

SPC Scott Geiszler, an Albany, OR native, leads Iraqi children in some physical training by introducing them to the overhead arm clap in the East Anbar Province June 6. Geiszler serves as an infantryman with B/52 Infantry Regiment "Hammer." (Photo by SGT Brad Willeford)

SGT Randy Stephens, a Dayton, TX native, searches a car at a vehicle checkpoint on June 8 in the Abu Ghuraib Qada. Stephens is an infantry team leader with C/1-27 Infantry Regiment "Wolfhounds." (Photo by SPC Robert Pegg)

MANRIQUE
BROWN
CASCINO
مركز شرطة التاجي
AL - TAJI POLICE STATION
RYAN

PFC Stephen Miller, of Batesville, AR, cleans his rifle while conducting refit operations at Camp Taji on June 16. Miller is an infantryman with A/1-27 Infantry Regiment "Wolfhounds." (Courtesy Photo)

SSG Joseph Perminas, an Antioch, IL native, greets a boy in the East Anbar Province during a patrol to help establish community relations in this area June 16. Perminas serves as an infantry squad leader with B/52 Infantry Regiment "Hammer." (Photo by SGT Brad Willeford)

CPT Lucas Yoho, a native of Long Beach, CA, discusses plans of a possible location of a police station with Iraqi Police officials in an attempt to bring more police to the Taji Qada, northwest of Baghdad, June 16. Yoho commands A/1-27 Infantry Regiment "Wolfhounds." (Photo by SSG JB Jaso III)

Soldier continues family legacy

By Pfc. John Ahn
1st Bn., 27th Inf. Regt., 2nd SBCT, 25th Inf. Div., MND-B

SFC Kenneth Palumbo wanted to do something exciting, serve his country and not walk too much. These requirements are what the San Diego native told his recruiter before joining the Army. That conversation led him to enlist as an armor Soldier.

His enlistment began another generation of military service for his family.

Today, Palumbo serves as a mobile gun system platoon sergeant assigned to C/1-27 Infantry Regiment "Wolfhounds."

The seeds of service were planted by his grandfather, who served in the United States Navy during World War II, and his stepfather, who served in Vietnam.

"My stepfather has shrapnel in his legs from grenades and landmines," said Palumbo. "He's one of the toughest guys I know."

With these influences, Palumbo began his Army career as an armor Soldier, and is now responsible for 30 armor and infantry Soldiers in his unit.

Before deploying, Palumbo was a recruiter, where he inspired more than 60 regular people to become Soldiers. He then did his part to prepare other units to deploy in support of Operation Iraqi Freedom and Operation Enduring Freedom as a training noncommissioned officer at the National Training Center, Fort Irwin, CA. After those missions, it was finally time for him to deploy.

"I was ready to get over here," he said. "I was finally getting my chance to do my part."

He and his team now conduct regular patrols, search for enemy caches, conduct village assessments, and engage with the residents of the area. Palumbo's commander said he is a good leader who is always concerned about his Soldiers.

"(SFC) Palumbo always puts the needs of his Soldiers above his own," said CPT David Uthlaut, a native of Charleston, SC. "He is a humble and approachable leader who has demonstrated on numerous occasions that he is willing to expend significant personal effort to address his Soldiers' concerns."

In March, he re-enlisted for his final time, confirming that he is a career Soldier and will remain in the Army for the next eight years. Palumbo said he hopes those eight years will allow him to reach all of his goals and continue to be an inspiration to his Soldiers.

SFC Kenneth Palumbo, a native of San Diego, is a platoon sergeant assigned to C/1-27 Infantry Regiment "Wolfhounds." (Photo by SGT Brad Willeford)

Shrevport, LA native, SSG Derrick Hardy coordinates for additional supplies needed on June 17 from Camp Taji. Hardy is the fire support non-commissioned officer for the 1-27 Infantry Regiment "Wolfhounds." (Photo by CPT Joshua May)

SSG James Hall, a native of Longview, TX, pets a calf during a dismounted patrol through the New Taji Nahia, located northwest of Baghdad on June 18. Hall is a reconnaissance scout team leader assigned to HHC/1-27 Infantry Regiment "Wolfhounds." (Photo by CPT Chad Chenoweth)

San Diego native, PFC Eric Gonzalez completes his shift in the battalion's communications office at Camp Taji on June 18, assisting in ensuring communications are not interrupted. Gonzalez is a communications specialist assigned to HHC/1-27 Infantry Regiment "Wolfhounds." (Photo by CPT Joshua May)

SPC Michael Tull, a Woodbury, NJ native, uncovers several 120mm mortars at an illegal weapons cache site on June 18 in the Abu Ghuraib Qada. Tull serves as an infantryman with C/1-27 Infantry Regiment "Wolfhounds." (Photo by SGT Mitchell Hassler)

SGT Denny Toussaint, a native of Brooklyn, NY, distributes hygiene items to children in a remote village in the Taji Qada June 18. Toussaint is an infantry team leader assigned to B/1-27 Infantry Regiment "Wolfhounds." (Photo by SSG JB Jaso III)

PFC Kirill Grinchenko, a native of Brooklyn, NY, talks to children while on patrol the Taji Qada June 18. Grinchenko is an infantryman assigned to B/52 Infantry Regiment "Hammer."
(Photo by SSG JB Jaso III)

SFC Carlito Dizon, a San Diego native, hands a soccer ball to a woman in the Taji Qada June 18. Dizon is an infantry platoon sergeant assigned to B/1-27 Infantry Regiment "Wolfhounds."
(Photo by SSG JB Jaso III)

PFC Kirill Grinchenko, a Brooklyn, NY native is followed by a group of children while on patrol in the Taji Qada June 18. Grinchenko is an infantryman assigned to B/52 Infantry Regiment "Hammer." (Photo by SSG JB Jaso III)

In his first week in Iraq, PVT Jesse Huckleberry, a native of Bowling Green, OH, greets a child while on patrol in the Taji Qada, northwest of Baghdad, June 18. Huckleberry is an infantryman assigned to B/1-27 Infantry Regiment "Wolfhounds."
(Photo by SSG JB Jaso III)

SGT Wiley Baker, a native of Warrensburg, MO, walks hand in hand with a child while on patrol in the Taji Qada June 18. Baker is an infantry team leader assigned to B/1-27 Infantry Regiment "Wolfhounds." (Photo by SSG JB Jaso III)

MAJ Patrick Aspland, a Fort Ann, NY native observes marketplace improvements with a sheik and businessman in the Taji Qada June 19. Aspland is the executive officer for 1-27 Infantry Regiment "Wolfhounds." (Photo by SSG JB Jaso III)

1LT Roy Thomas, a Knoxville, TN native; and SSG Joseph Dearth, a Lima, OH native, talk with several children in the Taji Qada June 21. Dearth is an infantry squad leader with B/1-27. and Thomas is an infantry platoon leader with B/52. (Photo by SGT Brad Willeford)

SPC Joseph Santalucia, a Glens Falls, NY native, takes time to spread a little joy, not only to the people of the Taji Qada with whom he works to help on a daily basis, but also to one lucky cow June 21. Santalucia serves as an infantryman with B/52 Infantry Regiment "Hammer." (Photo by SGT Brad Willeford)

SPC Timothy Russell, a native of Yardley, PA, patrols through a marketplace in the Taji Qada ensuring the safety of residents conducting business June 19. Russell is a mortarman assigned to HHC/1-27 Infantry Regiment "Wolfhounds." (Photo by SSG JB Jaso III)

(Left) CPT Arnulfo Martinez, a native of San Antonio; LTC Richard "Flip" Wilson, a native of Brooklyn, NY; CPT Brian Sweigart, a native of San Antonio, and 1SG Timothy Jackson, a native of Groton, NY, prepare for a change of command ceremony at Camp Taji June 22. Martinez relinquished command of HHC/1-27 Infantry Regiment "Wolfhounds" to Sweigart.
(Photo by SSG JB Jaso III)

For the final time, CPT Arnulfo Martinez, a native of San Antonio, hands the company guidon to LTC Richard "Flip" Wilson, a native of Brooklyn, NY, as he officially relinquishes command of HHC/1-27 Infantry Regiment "Wolfhounds," at Camp Taji, northwest of Baghdad, June 22. Martinez was in command for more than 24-months. Wilson commands the "Wolfhounds." (Photo by SSG JB Jaso III)

CPT Arnulfo Martinez, a native of San Antonio delivers his farewell speech after relinquishing command of HHC/1-27 Infantry Regiment "Wolfhounds," at Camp Taji, northwest of Baghdad, June 22. Martinez was in command for more than 24 months and will assume a brigade-level staff position. (Photo by SSG JB Jaso III)

CPT John Goldsworthy, of Rosalia, WA, meets with a Sons of Iraq contact holder at Joint Security Station Hor Al Bosh on Jun. 23. Goldsworthy commands B/52nd Infantry Regiment "Hammer," attached to the 1-27 Infantry Regiment "Wolfhounds." (Courtesy Photo)

SPC John Pearson, a Kennesaw, GA native, enjoys time to play games with a child, in the Taji Qada located northwest of Baghdad, June 24. Pearson serves as an assist team leader with TPT 1624. (Photo by SPC Brian Pierce)

SPC James Simmons, a Jacksonville, FL native, searches for any suspicious areas in the Taji Qada during Operation Hammer Bonzai VII June 25. The operation was aimed at locating and removing enemy weapons caches in the area. Simmons serves as an infantryman with A/1-27 Infantry Regiment "Wolfhounds." (Photo by SPC Brian Pierce)

CPL Jacob Crowder, a Monticello, IN native, reviews the controls of a Stryker armored vehicles with one of his soldiers on June 28, in the Abu Ghuraib Qada. Crowder is an infantry team leader with C/1-27 Infantry Regiment "Wolfhounds." (Photo by SGT Mitchell Hassler)

SPC Gregory Kastner, a native of Excelsior, MN, provides security following a sheik support council meeting in the Taji Qada June 28. Kastner is a communications specialist assigned to HHC/1-27 Infantry Regiment "Wolfhounds." (Photo by SSG JB Jaso III)

SPC Eric Waddle takes a break as he observes Iraqi workers renovating a school in Al Awad, July 17. Waddle is a mortarman with HHC/1-27 Infantry Regiment "Wolfhounds." (Photo by SPC Daniel Herrera)

SGT James Gutierrez, of Amarillo, TX, pulls his weight during the platoon competitions celebrating the U.S. Independence Day at Joint Security Station Hor al Bosh on Jul. 1. Gutierrez is the CBRN NCO for B/52nd Infantry Regiment "Hammer," attached to the 1-27 Infantry Regiment "Wolfhounds." (Photo by CPT John Goldsworthy)

SGT Wiley Baker, a Warrensburg, MO native, works with a soldier from the 2nd Battalion, 31st Brigade, 8th Iraqi Army Division in the Taji Qada during Operation Hammer Bonzai VIII, a mission aimed at clearing the area, of illegal weapons, July 2. Baker serves as an infantry team leader with B/1-27 Infantry Regiment "Wolfhounds." (Photo by SGT Brad Willeford)

PFC William Gentry, an Austin, TX native, digs for buried weapons during Operation Hammer Bonzai VIII, a mission aimed at clearing the Taji Qada, northwest of Baghdad of enemy weapons caches on July 2. Gentry serve as an infantryman with B/1-27 Infantry Regiment "Wolfhounds." (Photo by SGT Brad Willeford)

SGT Wiley Baker, a Warrensburg, MO native, works with an Iraqi soldier from the 2nd Battalion, 31st Brigade, 8th Iraqi Army Division during Operation Hammer Bonzai VIII, in the Taji Qada July 2. The mission was aimed at clearing the area of weapons caches. Baker serves as an infantry team leader with B/1-27 Infantry Regiment "Wolfhounds." (Photo by SGT Brad Willeford)

SFC Carlito Dizon, a San Diego native, patrols down a dirt road ensuring the area is secure during Operation Hammer Bonzoi VIII, a mission aimed at locating enemy weapon caches in Taji Qada July 2. Dizon serves as an infantry platoon sergeant with B/1-27 Infantry Regiment "Wolfhounds." (Photo by SGT Brad Willeford)

LTC Grant Forrester, a native of Cedar Falls, IA, gives a lecture on trauma injuries to the head to the staff at the Taji Medical Clinic in the Taji Qada, northwest of Bagdad, July 2. Forrester is the battalion surgeon for 1-27 Infantry Regiment "Wolfhounds." (Photo by PFC John Ahn)

An Iraqi man is surrounded by children July 2 as he gives out Iraqi flags at a marketplace in the Taji Qada, northwest of Baghdad. (Photo by SSG JB Jaso III)

CPT Ryan Clay, a native of Durham, NC, thumb wrestles with a child July 2 in a Taji Qada marketplace, northwest of Baghdad. Clay is the mortar platoon leader for HHC/1-27 Infantry Regiment "Wolfhounds." (Photo by SSG JB Jaso III)

MAJ Patrick Aspland, a native of Fort Ann, NY, gives a stuffed animal to an Iraqi girl July 2 while on patrol in a marketplace in the Taji Qada, northwest of Baghdad. Aspland is the executive officer for 1-27 Infantry Regiment "Wolfhounds." (Photo by SSG JB Jaso III)

CPT Brian Sweigart, a native of San Antonio, talks with several children July 2 while on patrol in a marketplace in the Taji Qada, northwest of Baghdad. Sweigart is the commander, HHC/1-27 Infantry Regiment "Wolfhounds." (Photo by SSG JB Jaso III)

CPT Andrew Farina, a Franklin, NH native, describes events planned to Soldiers at Joint Security Station Sheik Amir in East Anbar July 4. The Soldiers celebrated Independence Day with platoon competitions. Farina commands B/1-27 Infantry Regiment "Wolfhounds." (Photo by SPC Brian Pierce)

SPC Gregory Greene, a native of Hamer, SC, bench presses at an Independence Day celebration on Joint Security Station Shiek Amir in East Anbar July 4. Greene is an infantryman assigned to B/1-27 Infantry Regiment "Wolfhounds." (Photo by SPC Brian Pierce)

SGT John White, a Melbourne, FL native, completes pull- ups during a day of competitions at Joint Security Station Shiek Amir in East Anbar July 4. White is a fire support Non-commissioned Officer for B/1-27 Infantry Regiment "Wolfhounds." (Photo by SPC Brian Pierce)

PFC Matthew Deagle, a Hardyville, VA native, flips a tire during Independence Day competitions at Joint Security Station Shiek Amir in East Anbar July 4, while his teammates encourage him to continue. Deagle is an infantryman assigned to B/1-27 Infantry Regiment "Wolfhounds." (Photo by SPC Brian Pierce)

Wolfhound joined Army for excitement

By SGT Brad Willeford
1st Bn., 27th Inf. Reg., 2nd SBCT, 25th Inf. Div., MND-B

PFC Tommy Smith, a Rome, GA, native, joined the U.S. Army in 2006. He said he joined the Infantry looking for excitement and adventure, because that is "where all the fun is at."

Smith is deployed with A/1-27 Infantry Regiment "Wolfhounds," and is having the adventure he always looked for.

Before joining the Army, Smith worked construction and in his spare time practiced his martial arts which he learned from his father. In high school Smith played football and spent his weekends working on his truck and hanging out with friends. But that sense of adventure was not there, that is when he decided to see what the Army had to offer.

His first sense of adventure came quickly after signing his enlistment contract. He took his first plane ride to Basic Combat Training at Fort Benning, GA. That was one of the many firsts for Smith in his military career
His next stop was Schofield Barracks, Hawaii.

While in Hawaii, Smith took advantage of all that Hawaii has to offer; spending his days hiking, snorkeling at shark's cove, and driving around the island.
When Smith received word of his imminent deployment to Iraq he took the news in stride.

"I knew full well that I would be going to Iraq," he said. He said the news was not a shock and he looked forward to the challenge.

On the flight to Kuwait, Smith said he felt a little anxious and full of anticipation as to what it would be like once he arrived in Iraq. Even though he had these thoughts he was able to take strength in knowing he knew his job, and knew it well.

"He is always ready for any mission, just says 'roger' and moves out. He is also very handy with tools, he chips in where ever he can," said SFC Raymond Zwicker, a native of Modesto, CA, and Smith's platoon sergeant.

In his first seven months in Iraq, Smith has been on over seventy-five missions, working as a member of the Personal Security Detail for the company commander.

As a part the PSD, it is Smith's job is to protect and provide security for the commander when he goes on missions, and out into the community to address the different situations that may arise.

Smith said he likes his job and the opportunity it gives him to have more interaction with the Iraqi population. "Most of the people are not here to do anything bad to us, I like knowing that I am helping them" Smith said.

He also said he enjoys being able to talk with the kids "They remind me of my cousins back home," Smith said after giving several children pieces of candy that he keeps in his pocket for them.

Since the company's arrival, they have been involved in numerous outreach programs in the Taji community.

The company operates in the Taji Qada, which is northwest of Baghdad.

"Things seem to be going great right now, but there is always room for improvement" Smith said. And that is the attitude that is helping the American soldiers win the hearts and minds, and improve the lives of the Iraqi people.

PFC Tommy Smith, a Rome, GA, native, checks the security of the Taji Qada, northwest of Baghdad, Feb. 20, along with Iraqi Army soldiers. Smith serves as an infantryman with A/1-27 Infantry Regiment "Wolfhounds." (Photo by SGT Brad Willeford)

SGT Jeremiah Harpster, an Austin, TX native, helps his Soldiers cross a canal while on a dismounted patrol through the Taji Qada July 5. Harpster serves as an infantry team leader with A/1-27 Infantry Regiment "Wolfhounds." (Photo by SGT Brad Willeford)

SSG Brandon Tulloch, a Marriottsville, MD native, meets a business owner while on patrol in the Taji Qada July 5. Tulloch serves as an infantry squad leader with B/52 Infantry Regiment "Hammer." (Photo by SGT Brad Willeford)

SPC Joseph Santa Lucia, a Glens Falls, NY native, talks with children during a patrol through the Taji Qada July 5, while assisting business owners in reestablishing economic stability. Santa Lucia serves as an infantryman with B/52 Infantry Regiment "Hammer." (Photo by SGT Brad Willeford)

SPC Matthew Lowery, a native of Salem, IL, gives candy to children while on patrol in the East Anbar Province July 7, Lowery and his team assesses essential services in the area. Lowery is a mortarman assigned to HHC/1-27 Infantry Regiment "Wolfhounds." (Photo by SSG JB Jaso III)

PFC Joseph Biando, a Chesapeake, VA native, provides security while military leaders discuss Government of Iraq grants available to business owners in the East Anbar Province July 7. Biando is a mortarman assigned to HHC/1-27 Infantry Regiment "Wolfhounds." (Photo by SSG JB Jaso III)

MAJ Patrick Aspland, a native of Fort Ann, NY, discusses grants available with a business owner that would assist in improving his store in the East Anbar Province July 7. Aspland is the executive officer for 1-27 Infantry Regiment "Wolfhounds." (Photo by SSG JB Jaso III)

With temperatures reaching over 115-degrees, SGT Jeffrey Coots, a Brockton, MA native, stays cool under an awning in a marketplace in the East Anbar Province July 7. Coots is a combat medic assigned to HHC/1-27 Infantry Regiment "Wolfhounds." (Photo by SSG JB Jaso III)

MAJ Patrick Aspland, a Fort Ann, NY native, is led to a compact water treatment unit, July 7, by CPL Michael Tejada, a Hastings, MN native, where potable water is produced for the residents in the East Anbar Province, northwest of Baghdad. Aspland is the executive officer for 1-27 Infantry Regiment "Wolfhounds," and Tejada is a mortarman assigned to B/1-27 Infantry Regiment "Wolfhounds." (Photo by SSG JB Jaso III)

PFC Joseph Biando, a native of Chesapeake, VA, provides security for military and Government of Iraq officials while inspecting a compact water treatment unit, July 7. The unit provides potable water to residents in the East Anbar Province, northwest of Baghdad. Biando is a mortarman assigned to HHC/1-27 Infantry Regiment "Wolfhounds." (Photo by SSG JB Jaso III)

A tribal sheik discusses the application process for citizens applying to become Iraqi Police officers with CPT Omololu Makinde, a Columbus, OH native, during an event where Government of Iraq officials screened potential future police officers at a Joint Security Station in the East Anbar Province July 8. Makinde is the Iraqi security forces coordinator for 1-27 Infantry Regiment "Wolfhounds." (Photo by SSG JB Jaso III)

SSG Kristopher Verrett, a Fort Myers, FL native, demonstrates pull ups to potential Iraqi Police officers at a Joint Security Station in the East Anbar Province July 8. Verrett is an infantry squad leader assigned to B/1-27 Infantry Regiment "Wolfhounds." (Photo by SSG JB Jaso III)

SPC Adam Brown, an Annandale, VA native, patrols through a neighborhood in the Taji Qada July 9, en route to a tribal sheik's house to discuss essential services and grants. Brown is an infantryman assigned to B/1-27 Infantry Regiment "Wolfhounds." (Photo by SSG JB Jaso III)

1LT Carlos Manrique, a native of St. Augustine, FL, discusses grants and essential services with a sheik in the Taji Qada July 9, ensuring that residents that live in the area are provided with much needed items. Manrique is an infantry platoon leader assigned to B/1-27 Infantry Regiment "Wolfhounds." (Photo by SSG JB Jaso III)

SPC Jonathan Martin, a Salem, OR native, departs a sheik's home after discussing essential services, security, and business grants while enjoying Iraqi cuisine in the Taji Qada July 9. Martin is an infantryman assigned to B/1-27 Infantry Regiment "Wolfhounds." (Photo by SSG JB Jaso III)

SFC Benjamin Chapman, of Barboursville, WV, maintains security at the entrance of the building where detainees will be released, July 9. Chapman is the platoon sergeant for the reconnaissance platoon assigned to HHC/1-27 Infantry Regiment "Wolfhounds." (Photo by PFC John Ahn)

SGT Denny Toussaint, a native of Brooklyn, NY, gives a visor and Iraqi flags to a boy in a Taji Qada marketplace July 9. Toussaint is an infantry team leader assigned to B/1-27 Infantry Regiment "Wolfhounds." (Photo by SSG JB Jaso III)

1LT Carlos Manrique, a St. Augustine, FL native, begins to enjoy lunch at a sheik's home after discussing business grants and much-needed essential services in the Taji Qada July 9. Manrique is an infantry platoon leader assigned to B/1-27 Infantry Regiment "Wolfhounds." (Photo by SSG JB Jaso III)

SPC Jonathan Martin, a native of Salem, OR, talks to children inside a Taji Qada marketplace, northwest of Bagdad, July 9, while military leaders discussed grants available to business owners to improve their shops. Martin is an infantryman assigned to B/1-27 Infantry Regiment "Wolfhounds." (Photo by SSG JB Jaso III)

PVT Joshua Cole, a Clarkston, MI native, talks with children during Operation Borzoi Bonzoi X, a mission aimed at clearing portions of the East Anbar Province of enemy weapons while also working to provide essential services for the people of this area, July 10. Cole serves as an infantryman with B/1-27 Infantry Regiment "Wolfhounds." (Photo by SPC Brian Pierce)

COL Todd McCaffrey talks with SPC Charles Joseph, a Kerrville, TX native, about his activities while on patrol earlier in the day, July 10, in the Taji Qada, northwest of Baghdad. Joseph serves as an infantryman with A/1-27 Infantry Regiment "Wolfhounds.".
(Photo by SGT Brad Willeford)

1LT Christopher Smith, a native of Daniel Island, SC, listens to tribal sheiks share their security and essential service concerns after a sheik support council meeting held in the Taji Qada July 12. Smith is the targeting officer for 1-27 Infantry Regiment "Wolfhounds" (Photo by SSG JB Jaso III)

SGT Timothy Miller, a Circleville, OH native, talks with several boys during a patrol through the Taji Qada July 12. Miller serves as an infantry team leader with A/1-27 Infantry Regiment "Wolfhounds."
(Photo by SGT Brad Willeford)

CPT Darrell Tran, a Bridgeport, CT native, addresses the Soldiers of his company for the first time during his change of command ceremony in the Taji Qada July 10. Tran serves as commander of B/52 Infantry Regiment "Hammer." (Photo by PFC John Ahn)

Soldiers of B/52 Infantry Regiment "Hammer" render their salute during a change of command ceremony in the Taji Qada July 10. (Photo by PFC John Ahn)

1SG Shiloh Butterworth, a Lebanon, OR native, holds the company's guidon during a change of command ceremony in the Taji Qada July 10. Butterworth serves as the first sergeant of B/52 Infantry Regiment "Hammer." (Photo by PFC John Ahn)

1LT Tommy Ryan, of Seminole, FL, holds up a young boy during a patrol through the Taji Qada July 12. Ryan serves as an infantry platoon leader with A/1-27 Infantry Regiment "Wolfhounds."
(Photo by SGT Brad Willeford)

SPC Allen Post, of San Antonio, patrols through a marketplace on July 17, while leaders assessed businesses that received grants to improve their stores from the Government of Iraq. Post is a combat medic with HHC/1-27 Infantry Regiment "Wolfhounds." (Photo by SSG JB Jaso III)

Children swarm SPC Timothy Russell, a Yardley, PA native when he tries to leave a marketplace in the Taji Qada July 17. Russell is a mortarman assigned to HHC/1-27 Infantry Regiment "Wolfhounds."
(Photo by SSG JB Jaso III)

SGT Eberardo Solis, a Mission, TX native, talks with a boy during a patrol through markets in the Taji Qada July 12. Solis serves as an infantry team leader with A/1-27 Infantry Regiment "Wolfhounds."
(Photo by SGT Brad Willeford)

PVT Richard Martinez, a Mission, TX native, uses a metal detector in search of enemy weapon caches during Operation Hammer Bonzai IX in the Taji Qada July 18. Martinez is an infantryman assigned to A/1-27 Infantry Regiment "Wolfhounds."
(Photo by SPC Brian Pierce)

SSG Dusty Lamborn, a Salt Lake City native, digs into the ground in extreme heat during Operation Hammer Bonzai IX, July 18. The purpose of the mission was to rid the Taji Qada from enemy weapon caches. Lamborn is an infantry squad leader assigned to A/1-27 Infantry Regiment "Wolfhounds." (Photo by SPC Brian Pierce)

SSG Donald Fabian, a Miami native, gives an Iraqi child candy during Operation Hammer Bonzai IX in the Taji Qada July 18. Fabian is a psychological operations team leader for the 1620th Tactical Psychological Operations Detachment, 312th Tactical Physiological Operations Company. (Photo by SPC Brian Pierce)

SPC Paul Piertz, a Leesville, LA native, searches people during a security stop on July 18 in the Abu Ghuraib Qada. Piertz serves as an infantryman with C/1-27 Infantry Regiment "Wolfhounds."
(Photo by SPC Robert Pegg)

SPC John Pearson, a Kennesaw, GA native, talks with an employee of a water treatment plant in the East Anbar Province, northwest of Baghdad during an assessment of the facility July 22. Pearson serves as an assistant team leader for the 1620th Tactical Psychological Operations Detachment, 312th Tactical Physiological Operations Company. (Photo by SGT Brad Willeford)

SFC Ernest Oberst, a Gladstone, MI native, teaches Iraqi children how to make the peace sign during Operation Hammer Bonzai XI, an operation focused at locating and destroying terrorist weapon caches in the Taji Qada July 22. Oberst serves as an infantry platoon sergeant in A/1-27 Infantry Regiment "Wolfhounds."
(Photo by SPC Brian Pierce)

Wolfhounds, GoI work to rebuild education

By SSG JB Jaso III
1st Bn., 27th Inf. Regt., 2nd SBCT, 25th Inf. Div., MND-B

The schoolhouse is run down. Water leaks through numerous cracks in the roof and air conditioners do not operate, effectively causing temperatures inside the building to climb to unbearable lengths.

The walls covered with mold, no doors to the classrooms, broken windows, and cracks in the walls large enough to see the surrounding farmlands.

Despite all this, approximately 600 students at the Al Abbas Primary School, located in northwest Baghdad's remote New Taji Nahia, continued to study and garner their education despite these difficult conditions.

CPT Shaun Wheelwright, a Pinetop, AZ, native, first visited the school in February and said he was saddened to see the conditions of the school. Wheelwright, the former executive officer for A/1-27 Infantry Regiment "Wolfhounds," saw that more than half of their desks were broken, without desktops or seats, and each desk was used for six or more students.

Wheelwright said he wanted to help in any way that he could.

That help began after he told his wife, Keri Wheelwright, about the poor conditions. He said she was also saddened to hear about the conditions and decided to help in anyway she could. She then contacted a Hawaiian-based company, Oils of Aloha, for assistance.

The first donation of school supplies reached the students April 26. Wheelwright led the "Wolfhounds" and Iraqi Army soldiers to the school to hand out the much-needed supplies which were sent by Oils of Aloha.

The Hawaiian-based company sent more than 20 large boxes of supplies, which included pencils, paper, notebooks, rulers, and more.

"We are very happy and grateful," said Braheem Ahmed Hamid, the school's headmaster, after receiving the school supplies April 26. "We are always in short supply of school supplies."

Wheelwright knew the supplies were a start, but it would not be enough. Immediately following that visit, the Wolfhound leadership worked with the Government of Iraq and applied for funding for reconstruction of the rickety schoolhouse.

The project was approved utilizing Iraqi Commanders Emergency Relief Program and Government of Iraq funding resulting in over $260,000 rewarded in June.

"The Al Abbas school was the least functioning school in the area," said CPT Brian Sweigart, a San Antonio, TX, native. Sweigart was the commander for A Co., and his unit was responsible for the area which the school was in. "It looked miserable. Mud was everywhere and it looked like it was falling apart. It was a very poor study environment."

The poor educational environment did not deter the children from furthering their education.

"The kids wanted an education, and they had no other choice," Sweigart said. "Three schools in the area were destroyed by terrorists some time ago, and now the Al Abbas school is the only school in the area."

Sweigart relinquished command of A Co. in April, then assumed command HHC/1-27 Inf. Regt., and took the lead of the battalion's embedded Local Reconstruction Team. He is now responsible for acquiring funds, and recommending projects such as the Al Abbas School.

One of his new duties includes inspecting the progress of the reconstruction projects. While traveling to the Al Abbas school to see the finished project June 17, Sweigart rode in the back of a Stryker armored vehicle with a smile on his face, looking eager to see the school. As the Stryker ramp was lowering, Sweigart yelled, "I'm excited!"

Sweigart wasn't the only one excited.

Sheik Yasin Muhammad Hussein al Halbusi, and Sheik Noori Athiyah Albu Regehba, the tribal sheiks for the area, where present to meet Sweigart and the "Wolfhounds" to walk through the school.

Sweigart, who hasn't seen the school since April, walked through the newly refurbished school with the sheiks. The courtyard and classrooms that were previously muddy are now cemented. The cracks in the walls were fixed, and the school was resurfaced and painted.

Artwork adorned the exterior of the classrooms, including a map of the region. Environmental control units were installed, desks repaired, and restrooms built. The school looked new.

"I'm impressed," Sweigart said as he walked through what was previously the worse school in the area. "They did a good job (reconstructing the school). It is above standard."

Muhammad couldn't stop thanking the "Wolfhounds" for their efforts in focusing on this school. "Children from three villages attend this school, as this is the only school in the area. It is serving all the Families and children in this area," he said.

"Kids are happy that the school is rebuilt," Noori said. "Kids use to be full of mud before. Now they can go to class with clean shoes."

The "Wolfhounds" are no stranger to this excitement over schools. They have already completed construction of six schools in the area, and are currently assisting in the construction of eight others.

"Education is the foundation for any country," stated CPT Mark Weber, a Montrose, PA, native, and effects coordinator with 1-27 Inf. Regt. "Providing an environment where the students are able to concentrate on their studies will develop the next set of leaders for Iraq."

CPT Brian Sweigart, a San Antonio, TX, native, inspects a classroom at the Al Abbas Primary School in northwest Baghdad's remote New Taji Nahia July 17 after final refurbishments were made the school. Sweigart is the commander of HHC/1-27 Infantry Regiment "Wolfhounds." (Photo by SSG JB Jaso III)

A student of the Al Abbas Primary School explains locations on a map that is painted on the outside of the newly refurbished school July 17. (Photo by SPC Daniel Herrera)

MAJ Michael Zernickow, a Concord, CA native, arrives at the 2nd Brigade, 1st Iraqi National Police Division compound to attend a meeting to discuss current security concerns in the Istaqlal Qada July 22. Zernickow is the operations officer for 1-27 Infantry Regiment "Wolfhounds." (Photo by PFC John Ahn)

1LT Erik Wiesehan, a Canby, OR native, listens to area leaders during a meeting to discuss how to better establish essential services for the East Anbar Province July 22. Wiesehan is the embedded Local Reconstruction Team's essential services coordinator, 1-27 Infantry Regiment "Wolfhounds." (Photo by SGT Brad Willeford)

SPC Lee Arroyo, a Bronx, NY native, is able to keep a smile on his face in the extreme heat of the day while assisting to establish essential services in the East Anbar Province, northwest of Baghdad July 22. Arroyo serves as a mortarman with HHC/1-27 Infantry Regiment "Wolfhounds." (Photo by SGT Brad Willeford)

PFC Dwayne Haynes, a Montgomery, AL native, assesses a baby that is ill in the Istaqlal Qada, northeast of Baghdad, July 27. Haynes is a combat medic for B/52 Infantry Regiment "Hammer." (Photo by SSG JB Jaso III)

MAJ Michael Zernickow, a Concord, CA native, listens during a meeting with Istaqlal Qada leaders on July 24, to find out in what ways he can better assist the people of Iraq. Zernickow serves as the operations officer for 1-27 Infantry Regiment "Wolfhounds."
(Photo by SGT Brad Willeford)

CPT Mark Awad (left), of Union, NJ; and CPT Omololu Makinde, of Columbus, OH, enjoy a meal after a meeting with Istaqlal Qada leaders to discuss the improvement of this area, July 24. Awad serves as an intelligence officer, while Makinde serves as the Iraqi Security Forces coordinator for 1-27 Infantry Regiment "Wolfhounds." (Photo by SGT Brad Willeford)

CPT Lucas Yoho, a Long Brach, CA native, explains security measures of a Joint Security Station in the Istaqlal Qada to BG William Grimsley, deputy commanding general, Multi-National Division – Baghdad, July 25. Yoho commands A/1-27 Infantry Regiment "Wolfhounds." (Photo by SGT Brad Willeford)

Wolfhounds, GoI bring water back

By SSG JB Jaso III
1-27 Infantry Regiment "Wolfhounds"

Just over a year ago, residents in the East Anbar Province had potable drinking water, but terrorists destroyed the main source of potable water in an attack on a nearby school. "The CWTU (Compact Water Treatment Unit) was originally built by the Army Corps of Engineers," said CPT Brian Sweigart, a San Antonio native, and embedded Local Reconstruction Team leader for 1-27 Infantry Regiment "Wolfhounds." "The day it opened-up in June 2007, terrorists blew-up the school next to it, which caused severe damage (to the CWTU)."

The $2 million project was disabled, cutting off water supply to more than 10,000 residents. Residents were left without the most essential human need. When the Wolfhounds arrived in January, Sweigart's embedded Local Reconstruction Team quickly identified the CWTU as one of their main priorities and sought out funding options to bring water back to the people. Iraqi Commander's Emergency Relief Program funding was awarded to not only repair the damaged CWTU, but to extend the water network to over 12,000 residents.

"The further away from the main line residents are, the harder it is to get potable drinking water," Sweigart said on the importance of the extension of the water network. "Once the CWTU is working at its full capacity, it will be able to provide potable water to over 12,000 people a day."

Each pump will generate 200-liters of potable water an hour, said 1LT Erik Wiesenhan, a Canby, OR native, and the essential services coordinator for the Wolfhound eLRT. "There are two pumps, and once they are fully operational they will be pumping 400-liters of water an hour together. That will allow us to extend the water to additional villages in the area," he said.

Wolfhounds toured the CWTU, July 22, to check on repairs made in the first month of the project. Water is now pumping into the storage tanks, and Government of Iraq officials will visit the site and inspect progress, and the project will continue. Half of the CWTU is now operational and workers are now laying the pipes to extend the network. Funding is now approved for the second part of the CWTU to be fixed.

"Over 12,000 human beings were not receiving water because of a war that was going on around them," Sweigart said. "There was destruction and things being blown-up around them, and people are suffering because of it. It is the basic human nature to help people like this. We've been able to provide a sustainable service that the Government and people of Iraq want and need."

CPT Brian Sweigart, a native of San Antonio, discusses the use of a compact water treatment unit that supplies water to more than 5,000 residents in Baghdad's Adhamiyah Security District on Aug. 19. Sweigart commands HHC/1-27 "Wolfhounds." (Photo by SSG JB Jaso III)

CPL Michael Tejada, a native of Hastings, MN, is greeted by a young Iraqi man while patrolling a village in the Istaqlal Qada, northeast of Baghdad, July 27. Tejada is a mortar section leader for B/1-27 Infantry Regiment "Wolfhounds."
(Photo by SSG JB Jaso III)

SSG Rudy Mello, a Houston native, flexes his muscles for an Iraqi child while visiting an Istaqlal Qada village, northeast of Baghdad, July 27. Mello is a squad leader for B/1-27 Infantry Regiment "Wolfhounds." (Photo by SSG JB Jaso III)

SPC David Sorensen, a Missouri City, TX native, plays with an Iraqi child dressed in camouflage, while patrolling the Istaqlal Qada, northeast of Baghdad, July 27. Sorensen is an infantryman with B/1-27 Infantry Regiment "Wolfhounds." (Photo by SSG JB Jaso III)

PVT Ryan Winklemann, a Huntsville, AL native, talks with children in the Istaqlal Qada, northeast of Baghdad, while patrolling the area, July 27. Winklemann is a mortarman with B/1-27 Infantry Regiment "Wolfhounds." (Photo by SSG JB Jaso III)

SPC Nicholas Snyder, a Martinsburg, PA native, speaks with a merchant at a marketplace during a patrol in the Istaqlal Qada, northeast of Baghdad, July 28. Snyder is a human intelligence analyst assigned to A/1-27 Infantry Regiment "Wolfhounds." (Photo by PFC John Ahn)

SSG Kalei Akana, a Honolulu native, talks into the radio during a marketplace patrol in the Istaqlal Qada, northeast of Baghdad, July 28. Akana is a squad leader assigned to B/52 Infantry Regiment "Hammer." (Photo by PFC John Ahn)

SPC Jose Maganda, a Riverside, CA, native, maintains security at a vehicle checkpoint on July 28 in the Abu Ghuraib Qada. Maganda serves as an infantryman with C/1-27 Infantry Regiment "Wolfhounds." (Photo by SGT Mitchell Hassler)

SGT Mark Davenport, a Wimberly, TX native, listens to a doctor at a hospital July 29 in the Istaqlal Qada, northeast of Baghdad, which was destroyed by terrorists in 2003. They spoke about what is needed to greater serve the residents in the area. Davenport is a combat medic assigned to A/1-27 Infantry Regiment "Wolfhounds." (Photo by SSG JB Jaso III)

CPT Kenneth Smith, a Phoenix native, walks through the rubble of a damaged hospital July 29 in the Istaqlal Qada, northeast of Baghdad. Smith is assessing what is needed to reestablish the medical facility. Smith is the physician assistant for 1-27 Infantry Regiment "Wolfhounds." (Photo by SSG JB Jaso III)

SGT Robert Lacy, of Bandera, TX, joins children fishing while assessing water treatment facilities, which provides potable water to residents in the Istaqlal Qada, northeast of Baghdad, July 31. Lacy is a mortar section leader assigned to HHC/1-27 Infantry Regiment "Wolfhounds." (Photo by SSG JB Jaso III)

1LT Erik Wiesehan, a Canby, OR native, and Dr. Natik Joudi, an Auckland, New Zealand native, examine a canal July 31, which is part of the Istaqlal Qada water network and is northeast of Baghdad. Wiesehan is the essential services coordinator for 1-27 Infantry Regiment "Wolfhounds," and Joudi is an essential services advisor for the 2nd SBCT. (Photo by SSG JB Jaso III)

After taking a picture of a group of young Iraqi men at a cooperative medical engagement in Hor Al Bosh, Iraq, U.S. Army SPC Charles Joseph, 1-27th Infantry Regiment "Wolfhounds," shows them the picture, July 30, 2008. (Photo by SPC Daniel Herrera)

Grinds into Soldier of the Quarter Cook is the Battalion's best Soldier

By SGT Brad Willeford
1st Bn., 27th Inf. Regt., 2nd SBCT, 25th Inf. Div., MND-B

Army infantryman SPC Bradley Cook, a native of Port Heron, MI, joined the U.S. Army on Jan. 20, 2006. Cook has come a long way from his days of skate boarding, playing football, and working at a local home improvement store. Cook, a former skater turned infantryman is now serving with C/1-27 Infantry Regiment "Wolfhounds."

After completing the basic infantryman's course at Fort Benning, GA, Cook was assigned to the "Wolfhounds," at Schofield Barracks, Hawaii. When Cook decided to be an infantryman, he knew he would be going to Iraq.

"When I told the recruiter what I wanted to do, he told me there is no doubt that I would be coming to Iraq" Cook said.

That did not deter him. He attributed the upcoming events as a valuable learning experience for his future, which he intends to pursue a career on a SWAT team in his local law enforcement.

When Cook was told that he would be leaving for Iraq, he was a little nervous and anxious, but excited. He was ready for any challenges and felt confident in the training the unit conducted at Schofield Barracks, Pohakuloa Training Area on the Big Island of Hawaii, and his combined training at the National Training Center at Fort Irwin, CA.

Cook has been on daily missions in support of Operation Iraqi Freedom since arriving to Iraq in December. On his first mission, Cook said he felt a little apprehensive going out, "not knowing exactly what to expect," he said. "Everything I had learned came back to me, and I adjusted."

He was recently on a mission that lasted for over nine hours, visiting all the Joint Security Stations in the brigade's area of operations, and assisting his platoon become more familiar with the area. After returning from this patrol, Cook said he was exhausted, but found it rewarding knowing he had an opportunity to help others.

"I just love it, not knowing what the next day will hold," Cook stated about C Company, "Coldsteel," which he calls his home away from home. "I am addicted now."

Cook has been busy in his first seven months in Iraq. He was a member of the 2nd SBCT's tactical reserve where he was responsible for rapidly responding to significant activities in the brigade's operational environment in the Taji Qada, located northwest of Baghdad. He was also responsible for recovering downed aircraft and unmanned aerial vehicles in addition to aiding in the effort to rebuild Iraq's government and stabilize the local area. For two months in the spring, he was assisting the Iraqi Army soldiers in securing and stabilizing Sadr City. Now, he has been instrumental in assisting the Government of Iraq in providing security and essential services to the residents in the Abu Ghuraib Qada.

Even though Cook has had a very busy last seven months, his typical day calls for an early wake-up, conducting physical training, and spending time reviewing rehearsing and training for missions.

"He is a stellar performer who always puts his best foot forward on a consistent basis" said SSG Jason Menard, Cook's squad leader. "He's one of the best soldiers I have in my squad."

Cook cemented Menard's comments as he was recently awarded as the Task Force's Soldier of the Quarter. Not an easy task, outshining some of the best infantryman in the Army. He was drilled, by first sergeants in the task force, with questions testing his military knowledge.

Cook is looking forward to returning to Hawaii in roughly eight months. When asked what he would do first when his unit returned to Hawaii, he answered with a grin, "Well, the first thing I am going to do is skate." Cook continued to say that he misses Hawaii's weather and the beaches. He also plans to take a trip to Maui with his brother, Ryan, where they plan to relax and learn how to surf.

Cook is making the most of his time here in Iraq by remaining physically fit and training in preparation to attend ranger school, which he plans to attend upon his return. When asked what he misses most because of this deployment, he said, "My family and friends back home, and having my weekends to myself." Until the time comes to return to the United States, Cook is staying focused on the mission at hand.

SPC Bradley Cook, a native of Port Heron, MI, is an infantryman assigned to C/1-27 Infantry Regiment "Wolfhounds." (Photo by SPC Brian Pierce)

SPC Bradley Cook, a Port Heron, MI, native, provides security during a patrol in the Abu Ghuraib Qada, northwest of Baghdad on July 20. Cook, has been serving as an infantryman with C/1-27 Infantry Regiment "Wolfhounds." (Photo by SGT Daniel Mitchell, Jr.)

A remote controlled car fitted with a camera is deployed by Soldiers assigned to A/1-27 Infantry Regiment "Wolfhounds," to check out a suspected improvised explosive device in the Taji Qada, located northwest of Baghdad, July 6. (Courtesy Photo)

Clockwise from above: SSG JB Jaso III of Long Beach, CA, 1-27 Infantry Regiment, throws copies of the Baghdad Now newspaper to the local populace of Istaqlal, Iraq, from a Stryker armored vehicle, Aug. 16.

Soldiers from 1-27 Infantry Regiment "Wolfhounds" take a break from security detail to enjoy a cup of tea at a cooperative medical engagement in Hor Al Bosh, Iraq, July 30.

Iraqi and coalition forces pack up and head out after conducting a cooperative medical engagement in Hor Al Bosh, Iraq, July 30.

(Photos by SPC Daniel Herrera)

SPC Joshua Foster, a native of West Stockholm, NY, walks through a gate that still bears the old Bathe party symbol during a patrol in the Istaqlal Qada, Aug. 12. Foster serves as an mortarman with B/1-27 Infantry Regiment "Wolfhounds." (Photo by SGT Brad Willeford)

1LT Louis Cascino, a Windsor, NY native, greets a sheik in the Istaqlal Qada, northeast of Baghdad, Aug. 1. Cascino serves as a platoon leader with B/1-27 Infantry Regiment "Wolfhounds." (Photo by SGT Brad Willeford)

PFC Adam Scheid, a Detroit native, maintains security while conducting a dismounted patrol through the Abu Ghuraib Qada on Aug. 3. Scheid is an infantryman with C/1-27 Infantry Regiment "Wolfhounds." (Photo by SGT Mitchell Hassler)

PFC Joshua Paladini, a Springfield, OR native, receives crackers from a young boy in the Abu Ghuraid Qada on Aug. 3. Paladini has assisted the Iraqi Security Forces in keeping this area just north of Baghdad safe and secure. Paladini is an infantryman with C/1-27 Infantry Regiment "Wolfhounds." (Photo by SPC Robert Pegg)

SPC Jeffery Lavigne, a Linolakes, MN native, ensures security is kept during a reconciliation ceremony in the Abu Ghuraib Qada on Aug. 4. Lavigne serves as an infantryman with C/1-27 Infantry Regiment "Wolfhounds." (Photo by SGT Karl Schell)

PFC David Newkirk, a native of Newburgh, NY, gives a thumbs-up during a patrol through the Istaqlal Qada on Aug. 10. Newkirk is a sniper with HHC/1-27 Infantry Regiment "Wolfhounds." (Photo by CPT Chad Chenoweth)

SSG Rudy Mello, a Houston native, talks with several children during a patrol in the Istaqlal Qada Aug. 11. Mello serves as a squad leader with B/1-27 Infantry Regiment "Wolfhounds." (Photo by SGT Brad Willeford)

SGT Eric Burke, a native of Kettering, OH, patrols the Istaqlal Qada Aug. 11 to assist the residents in the continued security of the region located northeast of Baghdad. Burke serves as a team leader with B/1-27 Infantry Regiment "Wolfhounds." (Photo by SGT Brad Willeford)

SSG Joseph Perminas, a native of Antioch, IL, plays a few hands of "51" during a patrol in the Istaqlal Qada Aug. 12 while working to build friendships in this northeast Baghdad community. Perminas serves as a squad leader with B/52 Infantry Regiment, 1-27 Infantry Regiment "Wolfhounds." (Photo by SGT Brad Willeford)

SPC Scott Geiszler, a native of Albany, NY, plays with a young girl during a patrol in the Istaqlal Qada, Aug. 12. Geiszler is a favorite of the people in this northeast Baghdad region not only because of his dedication to his work, but also for the respect he shows to all of the people. Geiszler serves as a team leader with B/52 Infantry Regiment "Hammer." (Photo by SGT Brad Willeford)

1LT Louis Cascino, a native of Windsor, NY, walks with a leader of the Istaqlal Qada Aug. 12 to discuss issues troubling this northeast Baghdad region. Cascino serves as an infantry platoon leader with B/1-27 Infantry Regiment "Wolfhounds."
(Photo by SGT Brad Willeford)

SGT Nicolas Laduke, of Minneapolis, MN; SGT Berry Lebehn, of The Federated States of Micronesia (Pohnpei State); and Juan Alvarez, of Phoenix, AZ, put finishing touches to the newly installed Stryker hull protection kit (armor side skirts) on Aug. 12. These kits provided additional protection from roadside bombs to the Soldiers riding inside of the Stryker armored vehicle. Laduke and Lebehn are wheeled vehicle mechanics and Alvarez is a General Dynamics Land Systems contractor, all assigned to the 1-27 Infantry Regiment "Wolfhounds." (Photo by CW3 Danny K. Taylor)

LTC Richard "Flip" Wilson, a native of Brooklyn, NY, plays with the grandson of a sheik prior to a meeting, Aug. 17, to discuss security gains and essential services needed in the Istaqlal Qada. Wilson commands the 1-27 Infantry Regiment "Wolfhounds."
(Photo by SSG JB Jaso III)

CPT Andrew Farina, a native of Franklin, NH, listens intently to a sheik in the Istaqlal Qada, Aug. 17, as they discuss security improvements to protect the residents in the region northeast of Baghdad. Farina commands B/1-27 Infantry Regiment "Wolfhounds."
(Photo by SSG JB Jaso III)

Soldiers from the 1-27 Infantry Regiment "Wolfhounds," maintain security outside a mosque in the northern Adhamiyah security district on Aug. 18, allowing residents in this northeast Baghdad region to worship without fear of criminal or terrorist activity.
(Photo by SSG JB Jaso III)

PFC Albert Peralta (below), a native of Long Beach, CA enjoys lunch at a sheik's home in the Istaqlal Qada Aug. 17, with members of his platoon (above). Peralta is an infantryman assigned to B/1-27 Infantry Regiment "Wolfhounds." (Photos by SSG JB Jaso III)

PFC Albert Peralta, a native of Long Beach, CA, enjoys a traditional Iraqi lunch at a sheik's home in the Istaqlal Qada Aug. 17 while military leaders discussed security concerns. (Photo by SSG JB Jaso III)

SPC Timothy Russell, of Yardley, PA, patrols through an Adhamiya Security District marketplace on Aug. 19, ensuring that residents are able to conduct their shopping without fear of terrorist or criminal activity. Russell is a mortarman assigned to HHC/1-27 Infantry Regiment "Wolfhounds." (Photo by SSG JB Jaso III)

1LT John Busuego, an Ontario, CA native, patrols past children in Baghdad's Adhamiyah Security District on Aug. 19 and gains support of the populace in this northeast Baghdad region. Busuego is the economics development coordinator for the 1-27 Infantry Regiment "Wolfhounds." (Photo by SSG JB Jaso III)

SSG Jesus Robles, an El Paso, TX native, receives a greeting by a child while patrolling through Baghdad's Adhamiyah Security District on Aug. 19. Robles is an infantryman assigned to HHC/1-27 Infantry Regiment "Wolfhounds." (Photo by SSG JB Jaso III)

PFC Joseph Biando, a native of Chesapeake, VA, talks with children outside a recreation center in Baghdad's Adhamiyah Security District on Aug. 19. Biando is a mortarman assigned to HHC/1-27 Infantry Regiment "Wolfhounds." (Photo by SSG JB Jaso III)

SSG Jesse Okiyama, an Inarajan, Guam, native, is surrounded by children on Aug. 19, while assessing water filters that were installed in Baghdad's Adhamiyah Security District. Okiyama is the pay agent for the 1-27 Infantry Regiment "Wolfhounds."
(Photo by SSG J.B. Jaso III)

1LT John Busuego, a native of Ontario, CA, seizing the moment by photographing a small group of children in Baghdad's Adhamiyah Security District on Aug. 19. Busuego is the economics development coordinator for 1-27 Infantry Regiment "Wolfhounds."
(Photo by SSG JB Jaso III)

CPT Brian Sweigart, a San Antonio native, listens to improvements made to an Istaqlal Qada school Aug. 19. The school was damaged by an artillery attack more than a year ago. Sweigart commands HHC/1-27 Infantry Regiment "Wolfhounds."
(Photo by SSG J.B. Jaso III)

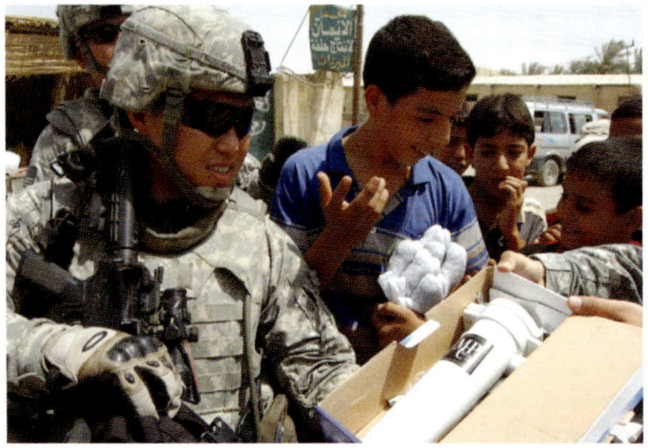

1LT John Busuego, an Ontario, CA, native, inspects a water filter that was supplied to 10,000 residents in Baghdad's Adhamiyah Security District by the Government of Iraq. Busuego is the economics development coordinator for the 1-27 Infantry Regiment "Wolfhounds." (Photo by SSG J.B. Jaso III)

CPT Sean Hollars, of Woodbine, MD walks with a community leader discussing the different reconstruction projects Aug. 19 that are occurring in this area northeast of Baghdad in the Rashidyah Nahia neighborhood. Hollars serves as the executive officer for B/1-27 Infantry Regiment "Wolfhounds." (Photo by SGT Brad Willeford)

Istaqlal Qada's Shiek Mohan Sajer Al Fayyad welcomes MAJ Patrick Aspland, a Fort Ann, NY native, to his home on Aug. 21, to discuss projects that will improve the quality of life for residents in this northeast Baghdad region. Aspland is the executive officer for the 1-27 Infantry Regiment "Wolfhounds."
(Photo by SSG JB Jaso III)

CPT Brian Sweigart (left), a San Antonio native, and 1LT John Busuego an Ontario, CA native, listen to a contractor discuss an initiative to provide drinking water to residents in the Istaqlal Qada on August 21. Sweigart and Busuego are members of the embedded local reconstruction team for the 1-27 Infantry Regiment "Wolfhounds." (Photo by SSG JB Jaso III)

CPT Andrew Farina, a Franklin, N.H., native, reviews a proposal for alternative means of power for the people in the Istaqlal Qada on Aug. 22, from a lack of electricity. Farina commands B/1-27 Infantry Regiment "Wolfhounds." (Photo by SGT Brad Willeford)

Wolfhounds conduct joint operations with Iraqi National Police

By PFC John Ahn
1st Bn., 27th Inf. Reg., 2nd SBCT, 25th Inf. Div., MND-B

In the pre-dawn light police officers of the 2nd Brigade, 1st Iraqi National Police Division and Soldiers of A/1-27 Infantry Regiment "Wolfhounds" linked up along a stretch of highway that connects the Diyala Province to Baghdad in the Istaqlal Qada, located northeast of Baghdad in order to search a city in the neighborhood of Hussaniya Nahia on Aug. 19. The densely populated areas of the Hussaniya Nahia is known as a supporter of roadside bombings, suicide attacks and cache storage sites.

CPT Lucas Yoho, a native of Long Beach, CA, commander of A Co. says, "the Iraqi National Police have done an outstanding job working hand in hand with coalition forces to help secure the Istaqlal Qada." The joint Wolfhound Stryker armored vehicles and the National Police armored trucks moved to Hussaniya Nahia. After quickly establishing the perimeter, NP officers and Soldiers, with a tactical attachment of bomb sniffing dogs and combat engineers began to search the area. While the search yielded no weapons or explosives, the operation was considered a success.

SGT Andrew Lambright, a Eugene, OR native, and infantry team leader with A Co. said, "The National Police integrated well with the mission objective. They were organized and tactically very sound. They worked well with Coalition Forces and were a huge success."

1LT Tommy Ryan, a Seminole, FL native, platoon leader with A Co., said, "Though it was the first time I worked with (Iraqi Policeman) 1LT Ehab, (2nd Bde., 1st Iraqi National Police Div.) He came extremely well prepared, proactive and had excellent knowledge of the area. He is the best Iraqi leader I've worked with."

(Top) Soldiers from A/1-27 Infantry Regiment "Wolfhounds" link up with officers from 2nd Brigade, 1st National Police Division, before sunrise on Aug. 20 on a dark deserted highway south of the Husayniyah Nahia.
(Bottom) During joint operations, Soldiers from A/1-27 Infantry Regiment "Wolfhounds" and officers from 2nd Brigade, 1st National Police Division search through the neighborhoods of Husayniyah Nahia.
(Photos by PFC John Ahn)

Soldiers of A/1-27 Infantry Regiment "Wolfhounds" take a knee during a cache searching operation in the Husayniyah Nahia, northeast of Baghdad, on Aug. 19. (Photo by PFC John Ahn)

A Soldier of A/1-27 Infantry Regiment "Wolfhounds" stands by to provide security as an Iraqi National Police officer searches the contents of an abandoned truck in the Husayniyah Nahia on Aug. 19. (Photo by PFC John Ahn)

SGT James Peptis of Colby, KS walks down a side road in the Istaqlal Qada on Aug. 19, while conducting a village assessment on essential services. Peptis serves as a infantry team leader with B/1-27 Infantry Regiment "Wolfhounds."
(Photo by SGT Brad Willeford)

SSG Patrick Santos, a native of Sinajana, Guam, plays with a young girl in the Istaqlal Qada on Aug.19, while talking with locals gathering the needs of the residents this northeast Baghdad community. Santos serves as a infantry squad leader with B/1-27 Infantry Regiment "Wolfhounds."
(Photo by SGT Brad Willeford)

CPT Mark Weber *(above)*, of Curtis Bay, MD, and SPC Joshua Foster *(below)*, of West Stockholm, NY, pull security during a patrol in the Istaqlal Qada on Aug. 22 to help provide security for the people northeast of Baghdad. Weber serves as the fire support officer for 1-27 Infantry Regiment "Wolfhounds", and Foster is an Infantryman with B/1-27 Infantry Regiment "Wolfhounds." (Photo by SGT Brad Willeford)

Solders with B/1-27 Infantry Regiment "Wolfhounds" take a knee while maintaining security during a patrol through palm groves in the Istaqlal Qada on Aug. 22, while working with the Iraqi government to keep the communities safe. (Photo by SGT Brad Willeford)

Wolfhound "Pup" excels in Iraq

By PFC John Ahn
1st Bn., 27th Inf. Regt., 2nd SBCT, 25th Inf. Div., MND-B

He is the youngest Soldier in his company and is affectionately known as the company "pup."

The Pup's real name is PFC Robert Hatala, and he is a reconnaissance scout assigned to HHC/1-27 Infantry Regiment "Wolfhounds."

Hatala joined the Army in July 2007, just three days after graduating high school in Ottawa, IL. His goal, he said, is to do his part and better his life and joining the Army is a stepping stone for future aspirations. Ultimately, he plans to pursue a career in special weapons and tactics law enforcement after using his college benefits from the Army.

In the meantime, he serves his country proudly.

"I knew it would come sooner or later," stated Hatala, in speaking about the inevitability of deploying to the warfront.

He rapidly deployed in support of Operation Iraqi Freedom in December 2007, where he is currently patrolling, securing and conducting reconnaissance missions in the Istaqlal Qada, to help ensure that the residents in this area, northeast of Baghdad, are free of criminal and terrorist activity.

Hatala has recently taken part in several successful missions that resulted in the capture of several of the brigade's and battalion's high-value individuals during a two-week surge to combat terrorist activity in the brigade area. Through the eight to 10 hour night missions, the 'pup' continued to keep his fellow Soldiers motivated through his jokes and uplifting personality.

"He is the most high energy person I have ever met. It doesn't matter how tired everyone else is. He always has energy," said SSG Michael Mucha, a Fairfax, VA, native, a section leader in the reconnaissance platoon.

The pace of missions is hard, but the reconnaissance platoon continues to stay motivated and detain terrorists who have hurt or killed innocent Iraqi people. Hatala rests a few hours and then continues to search for the enemy.

"It's a cool job," stated Hatala, "You go out on missions, look for the bad guys and provide accurate and timely information to other units to facilitate [future] operations."

Hatala was recognized by LTC Richard "Flip" Wilson, a native of Brooklyn, NY, the task force commander, for being the youngest "Havoc" soldier and was presented his commander's coin for his service.

He has a large family waiting and praying for his safe return home in 2009. His fiancé and sweetheart is raising their one-year-old son while he is deployed and is expecting twins in November. The young couple had planned to wed during his rest and recreation leave from combat, but the dates did not match up to include their Families. When he returns to Hawaii from combat operations, he said he plans to spend time with his children and get married as soon as possible.

PFC Robert Hatala, of Ottawa, IL, is a reconnaissance scout with HHC/1-27 Infantry Regiment "Wolfhounds." (Photo by PFC John Ahn)

SGT Billy Gibson *(above)* of Springfield, MO, grades the physical fitness screening of a hopeful Iraqi intending on joining the Iraqi Police on Aug. 23 in the Istaqlal Qada. Gibson is an infantryman assigned to HHC/1-27 Infantry Regiment "Wolfhounds." An Istaqlal Qada resident *(below)* participates in the Iraqi physical fitness screening.
(Photos by SSG JB Jaso III)

1LT Richard Robinson of Houston, TX identifies personnel that will be participating in the initial screening process to become an Iraqi Police officer on Aug. 23. Robinson is a police transition team leader with the 411th MP Company, 716th MP Battalion.
(Photo by SSG JB Jaso III)

MAJ Patrick Aspland, of Fort Ann, NY, patrols through the Boob al-Sham community, northeast of Baghdad, in search of children to give stuffed animals to on Aug. 25. Aspland is the executive officer for the 1-27 Infantry Regiment "Wolfhounds". (Photo by SSG JB Jaso III)

MAJ Patrick Aspland a Fort Ann, NY, native, gets a "five" from a young child while leaving the Boob al-Sham community, northeast of Baghdad, on Aug. 25. Aspland is the executive officer for the 1-27 Infantry Regiment "Wolfhounds." (Photo by SSG JB Jaso III)

SPC Eric Waddle, of Bedford, IN, patrols through a marketplace in Baghdad's Adhamiyah district on Aug. 25, with area children that aspire to one-day be Soldiers like him. Waddle is a mortarman in HHC/1-27 Infantry Regiment "Wolfhounds." (Photo by SSG JB Jaso III)

SSG Jesus Robles, of El Paso, TX, oversees the marketplace in Baghdad's Adhamiyah district, after patrolling this northeast Baghdad area, ensuring the continued safety of residents on Aug. 25. Robles is an infantryman in HHC/1-27 Infantry Regiment "Wolfhounds." (Photo by SSG JB Jaso III)

MP

SPC Christopher Tinsley, of Amarillo, TX, observes neighborhood activity in the Rashidyah Nahia, northeast of Baghdad, Aug. 31. Tinsley serves as a mortarman with B/1-27 Infantry Regiment "Wolfhounds." (Photo by SGT Brad Willeford)

MAJ James Moyes and SSG Herman Sims assist members of the Iraqi Security Forces in securing the the Istaqlal Qada on Aug. 29. Moyes and Sims are members of the National Police Transition Team, 1-27 Infantry Regiment "Wolfhounds." (Courtesy Photo)

SGT Ian Bittenbender, an Edmeston, NY native, takes a picture of a potential Iraqi Security Forces candidate in the Rashidyah Nahia, northeast of Baghdad, Aug. 31. Bittenbender serves as an infantryman with B/1-27 Infantry Regiment "Wolfhounds."
(Photo by SGT Brad Willeford)

SSG John White, a Melbourne, FL native, assists the Government of Iraq in screening possible Iraqi Security Forces candidates in the Rashidyah Nahia, northeast of Baghdad, Aug. 31. White is the fire support non-commissioned officer with B/1-27 Infantry Regiment "Wolfhounds."
(Photo by SGT Brad Willeford)

PFC Edgar Gutierrez, a Santa Rosa, CA, native, passes sunflowers in bloom while on patrol in the Taji Qada, northwest of Baghdad, June 16. Gutierrez is a mortarman assigned to HHC/1-27 Infantry Regiment "Wolfhounds." (Photo by SSG JB Jaso III)

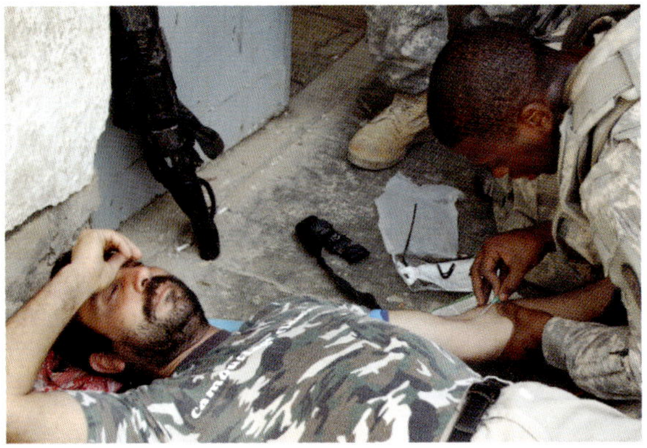

PFC Dwayne Haynes, of Montgomery, AL, aids a Rashidyah Nahia resident who is suffering from dehydration while on patrol in this Istaqlal Qada neighborhood, northeast of Baghdad Aug. 31. Haynes serves as a combat medic with B/52 Infantry Regiment, attached to the 1-27 Infantry Regiment "Wolfhounds."
(Photo by SGT Brad Willeford)

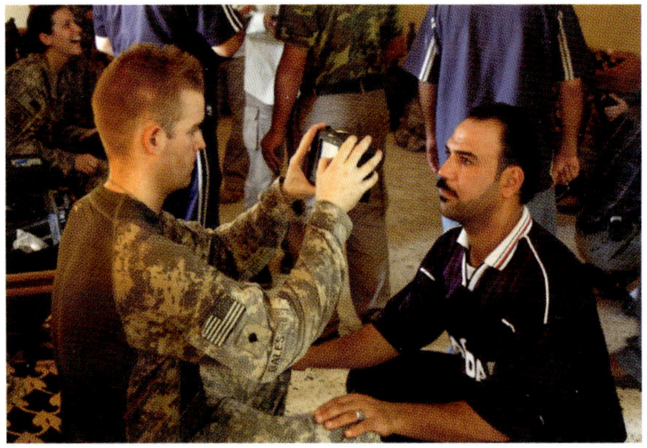

SPC Joshua Bales, an Incline Village, NV native, takes a photo of a Sons of Iraq (Abna al Iraq) member to enter him into a database of all SoI members in the Rashidyah Nahia, northeast of Baghdad, Sept. 3. Bales serves as an infantryman with B/1-27 Infantry Regiment "Wolfhounds." (Photo by SGT Brad Willeford)

PFC Joshua Cole, a Clarkston, MI native, takes a retinal scan of a Sons of Iraq (Abna al Iraq) member during a process to create an accurate database of all the SoI members in the Rashidiyah Nahia, northeast of Baghdad, Sept. 3. Cole serves as an infantryman with B/1-27 Infantry Regiment "Wolfhounds."
(Photo by SGT Brad Willeford)

Waipahu, HI native, SGT Wilburtigran Rayala; and Woodbury, NJ, native, SPC Michael Tull, maintain security while patrolling in the Dehna area, just northwest of Baghdad on Sept. 3. Rayala and Tull are both infantryman with C/1-27 Infantry Regiment "Wolfhounds."
(Photo by SGT Mitchell Hassler)

PV2 Justin Do, a Vancouver, WA, native, searches the Istaqlal Qada, northeast of Baghdad, Iraq, with Iraqi national police, in an effort to locate enemy weapon caches in support of Operation Gunslinger Bonzai XIV, Sept. 3. Do serves as an infantryman with A/1-27 Infantry Regiment "Wolfhounds." (Photo by SPC Brian Pierce)

Olean, NY, native, SGT Joseph Dunne; and Linolakes, MN, native SPC Jeffrey Lavigne maintain security back to back during a patrol in the Abu Ghuraib Qada on Sept. 4. Dunne and Lavigne are both infantryman with C/1-27 Infantry Regiment "Wolfhounds."
(Photo by SGT Karl Schell)

SSG Kristopher Verrett, of Fort Myers, FL, greets a young boy while on patrol in the Rashidyah Nahia on Sept. 4, assisting the Government of Iraq with security in this area northeast of Baghdad. Verrett serves as an infantry squad leader with B/1-27 Infantry Regiment "Wolfhounds." (Photo by SGT Brad Willeford)

CPT Peter Walther, of Mililani, HI, discusses future projects to restore water and electricity with Iraqi leaders in Hussaniyah town, northeast of Baghdad, Sept. 4. Plans are in the works to restore essential services to the people of the town. Walther is the executive officer for A/1-27 Infantry Regiment "Wolfhounds." (Photo by SPC Brian Pierce)

Soldiers from A/1-27 Infantry Regiment "Wolfhounds" ensure that residents are able to celebrate Ramadan without fear of criminal or terrorist activity during a patrol in the Husayniyah area of the Istaqlal Qada, northeast of Baghdad, Sept. 7. (Photo by SSG JB Jaso III)

Kennesaw, GA native, SPC John Pearson patrols through the Husseniyah Nahia on Sept. 7, ensuring the safety of residents in this northeast Baghdad region. Pearson is an assistant team leader with the 1620th Tactical Psychological Operations Detachment, 312th Tactical Psychological Operations Company. (Photo by SSG JB Jaso III)

SPC Bryan Davis, of Tonkawa, OK, approaches a vehicle for inspection of illegal weapons in Subak Sur community, northeast of Baghdad, Sept. 7. Davis is an infantryman assigned to A/1-27 Infantry Regiment "Wolfhounds." (Photo by SSG JB Jaso III)

PVT Justin Do, of Vancouver, WA, watches vehicles pass through a vehicle checkpoint designed to interdict enemy movement of illegal weapons in Subak Sur, northeast of Baghdad, Sept. 7. Do is an infantryman assigned to A/1-27 Infantry Regiment "Wolfhounds." (Photo by SSG JB Jaso III)

SPC John Pearson, a native of Kennesaw, GA, plays foosball with children in the Husseniyah Nahia, northeast of Baghdad, Sept. 8. Pearson is an assistant team leader with the 1620th Tactical Psychological Operations Detachment, 312th Tactical Psychological Operations Company, Multi-National Division – Baghdad. (Photo by SSG JB Jaso III)

PVT Richard Plasket, a Midlothian, TX native, maintains security in a Husseniyah marketplace, northeast of Baghdad, Sept. 8, allowing military leaders to discuss safety concerns with business owners. Plasket is an infantryman assigned to A/1-27 Infantry Regiment "Wolfhounds." (Photo by SSG JB Jaso III)

SFC Omar Ashley, of Corona, NY, helps Iraqi National Police search houses and empty lots Sept. 10 to find illegal weapon caches in Istaqlal Qada's Hussaniyah town, northeast of Baghdad, during Operation Gunslinger Bonzai XIV. Ashley serves as an infantry platoon sergeant with A/1-27 Infantry Regiment "Wolfhounds." (Photo by SPC Brian Pierce)

SGT Ryan Ruz, of San Diego, helps Iraqi National Police search houses and empty lots Sept. 10 in Istaqlal Qada's Hussaniyah town, northeast of Baghdad, during a combined operation locating and removing enemy weapon caches. Ruz serves as an infantryman with A/1-27 Infantry Regiment "Wolfhounds." (Photo by SPC Brian Pierce)

1LT Kenneth Jura, of Gulf Breeze, FL, turns his radio on before talking with Sons of Iaq (Abna al-Iraq) guards as they prepare to conduct a night patrol in the Rashidyah Nahia, northeast of Baghdad, Sept. 10. Jura is an infantry platoon leader with B/1-27 Infantry Regiment "Wolfhounds." (Photo by SGT Brad Willeford)

Pasadena, CA native, SPC Toribio Gurrola maintains security from inside an Abu Ghuraib Qada building on Sept. 10. Gurrola is an infantryman with C/1-27 Infantry Regiment "Wolfhounds." (Photo by SPC Robert Pegg)

Wolfhounds Plow Through Blizzard

By SSG JB Jaso III
1st Bn. 27th Inf. Regt., 2nd SBCT, 25th Inf. Div., MND-B

There is a blizzard hitting Husayniyah town, northeast of Baghdad.

This blizzard isn't cooling the summer temperatures, nor is it blowing snow. Rather Soldiers from A/1-27 Infantry Regiment "Wolfhounds"are weathering the 110-degree end of summer heat and severe dust storms in an effort to rid the area of terrorist activities as part of Operation Wolfhound Blizzard.

The operation began Sept. 1 and is an ongoing effort between the Wolfhounds and the 2nd Brigade, 1st Iraqi National Police Division, to ensure the safety of residents in the area.

Operation Blizzard increased a presence of NP and Wolfhounds in the Istaqlal Qada. The NP and Soldiers informed residents about criminals in the area and highlighted recent essential service gains.

"We are notifying residents of known terrorists that plan to return to the area from other locations," said 1LT Tommy Ryan, a native of Seminole, FL, and Wolfhound platoon leader. "The people are tired of living their life in fear, and they are calling us with information that we hope will lead to arrests."

While they receive information about the whereabouts of terrorists in the area, the Wolfhounds and NP continue to inform residents by disseminating wanted posters and flyers. "The other night, I was giving some men the wanted signs and they looked at the picture of the terrorist and immediately threw it on the ground," said an NP officer. "The men immediately began stomping the picture with the bottom of their shoes."

The officer said this reaction was a true indication people want to live their lives in peace and free from terrorists. The overwhelming support from the residents is allowing the INP and the Wolfhounds to continue to plow through the area in an effort to remove terrorists and ensure the safety of the residents.

SGT Andrew Lambright, of Eugene, OR, sticks a wanted poster on a wall in Husayniyah, northeast of Baghdad, Sept. 7 in support of Operation Wolfhound Blizzard. The operation began Sept. 1 and is an ongoing effort between the Wolfhounds and the 2nd Brigade, 1st Iraqi National Police Division to ensure the safety of residents in the area. Lambright is an infantry team leader assigned to A/1-27 Infantry Regiment "Wolfhounds." (Photo by SSG JB Jaso III)

MAJ Patrick Aspland, of Fort Ann, NY, points out an individual that is wanted for terrorist activity to a young man in Husayniyah, northeast of Baghdad, Sept. 11 during Operation Wolfhound Blizzard. Aspland is the executive officer for the 1-27 Infantry Regiment "Wolfhounds." (Photo by SSG JB Jaso III)

Brothers continue Family tradition of service

By SSG JB Jaso III
1-27 Infantry Regiment "Wolfhounds"

A Wolfhound Soldier marked a memorable birthday celebration on a date now associated with great despair and reflection. The date was even more significant because he was joined by his brother.

On Sept. 11, CPT Joshua May, a signal officer and native of Fall, PA, assigned to 1-27 Infantry Regiment "Wolfhounds," met with his older brother, CPT Joseph May, brigade surgeon for 2nd Brigade Combat Team, 1st Armor Division, Multi-National Division – Central.

The May brothers spent Sept. 11 catching up, and marked Patriots Day with a special Sept. 11 tribute concert at Forward Operating Base Hammer, southeast of Baghdad. Joseph plays keyboard in the band "12 Lead," along with volunteer musicians from 2nd Brigade Combat Team, 1st Armor Division.

Joshua helped support the band's audio-visual efforts, which included a video tribute to the victims of the Sept. 11, 2001 attacks and to the Soldiers who have fought in support of the Global War on Terror.

Army service runs deep in the May Family. Their father, COL James May III, is a senior Army chaplain in Heidelberg, Germany. He retires in October following 34 years of military service. COL May's children all followed in their father's footsteps into the Army.

COL May's eldest son, James May IV, served in the Army for six years, including a deployment to Iraq in 2005.

"When it came time for my unit to deploy, I was ready to continue a Family tradition," said Joseph, the fourth in the Family to serve in Iraq.

"It's an honor to be in the military, and I'm especially proud to follow in my brothers' footsteps here in support of Operation Iraqi Freedom," he said.

Following in his brother's footsteps is what brought the brothers together for a birthday reunion.

"It was a great reunion," Joshua noted. "It's not often that military Family members get to cross paths with each other, especially within a combat environment."

Ellen May, their mother, might have said it best: "I pray every day for my sons who are serving overseas, and I'm extremely proud of our military Family. My sons are true brothers in arms."

CPT Joshua May celebrates his birthday on Sept. 11, by visiting his brother, CPT Joseph May, both natives of Falls, PA, at Forward Operating Base Hammer, southeast of Baghdad. The May Family has a long tradition of war time service, which the two brothers continue today. CPT Joshua May is the battalion signal officer for 1-27 Infantry Regiment "Wolfhounds," while CPT Joseph May is the brigade surgeon for the 2nd Brigade Combat Team, 1st Armor Division, Multi-National Division – Central. (Courtesy photo)

SPC Jared Gabriel, of Phoenix, AZ, talks with shop owners in the Rashidyah Nahia, northeast of Baghdad, Sept. 11 to check on his business. Gabriel serves as an infantryman with B/1-27 Infantry Regiment "Wolfhounds." (Photo by SGT Brad Willeford)

SPC Jose Carrillo. of Ventura County, CA. prepares to remove a Mine Resistant Ambush Protected (MRAP) vehicle engine during maintenance on Sept. 12. Carrillo is a wheeled vehicle mechanic assigned to 1-27 Infantry Regiment "Wolfhounds."
(Photo by CW3 Danny K. Taylor)

SFC Omar Ashley, a Corona, NY native, listens to a business owner discuss his security concerns in the Husseniyah marketplace area, northeast of Baghdad, Sept. 8. Ashley is a platoon sergeant assigned to A/1-27 Infantry Regiment "Wolfhounds." (Photo by SSG JB Jaso III)

SPC Brandon Baricuatro, a Frisco, TX native, looks inside a store in the Husseniyah Nahia during an early morning patrol on Sept. 12, ensuring the continued security in this area, northeast of Baghdad. Baricuatro is a reconnaissance scout assigned to HHC/1-27 Infantry Regiment "Wolfhounds." (Photo by SSG JB Jaso III)

CPT Chad Chenoweth, a Hamilton, OH native, gives instructions to his Soldiers while conducting an early morning patrol through the Husseniyah Nahia, northeast of Baghdad, Sept. 12. Chenoweth is the reconnaissance platoon leader for HHC/1-27 Infantry Regiment "Wolfhounds." (Photo by SSG JB Jaso III)

SSG Bruce Hillier, a Twin Falls, ID native, pulls security alongside a police officer assigned to the 1st Battalion "Wolf," 2nd Brigade, 1st Iraqi National Police Division in the Rashidyah Nahia, northeast of Baghdad, Sept. 15. Hillier serves as an infantry squad leader with B/1-27 Infantry Regiment "Wolfhounds." (Photo by SGT Brad Willeford)

SSG Kelly Muna, a Santa Rita, Guam native, meets and talks with a group of people Sept. 19 in the Rashidyah Nahia, northeast of Baghdad. Muna serves as an infantry squad leader with B/1-27 Infantry Regiment "Wolfhounds." (Photo by SGT Brad Willeford)

SSG Bruce Hillier, of Twin Falls, ID, and SGT Patricio Mannings, of Lawton, OK, work to lay out a role of concertina wire during a village assessment in the Rashidyah Nahia, northeast of Baghdad, Sept. 20. Hillier serves as an infantry squad leader and Mannings serves as a infantry team leader with B/1-27 Infantry Regiment "Wolfhounds." (Photo by SGT Brad Willeford)

SSG Bruce Hillier, a Twin Falls, ID native, watches as his Soldiers establish security during a village assessment in the Rashidyah Nahia, northeast of Baghdad, Sept. 20. Hillier serves as an infantry squad leader with B/1 27 Infantry Regiment "Wolfhounds."
(Photo by SGT Brad Willeford)

SPC John Pearson, of Kennesaw, GA, carries a bundle of Baghdad Now newspapers to a business on Sept. 21. Pearson is an assistant team leader with the 1620th Tactical Psychological Operations Detachment, 312th Tactical Psychological Operations Company. (Photo by SSG JB Jaso III)

1LT Tommy Ryan, of Seminole, FL, arm wrestles with a young Husseniyah child on Sept. 21. Ryan is a platoon leader with A/1-27 Infantry Regiment "Wolfhounds." (Photo by SSG JB Jaso III)

Midlothian, TX native, PFC Richard Plasket maintains security overwatch from the top of a building in Baghdad's Adhamiyah security district on Sept. 22. Plasket is an infantryman with A/1-27 Infantry Regiment "Wolfhounds." (Photo by SSG JB Jaso III)

Phoenix native, SSG Louis Overson talks to a young girl in Baghdad's Adhamiyah security district on Sept. 22, while patrolling the area collecting atmospherics. Overson is an infantry squad leader with A/1-27 Infantry Regiment "Wolfhounds." (Photo by SSG JB Jaso III)

1LT Kenneth Jura, a Gulf Breeze, FL, native, plays with a boy in the Rashidyah Nahia neighborhood, northeast of Baghdad, Sept. 23. Jura, an infantry platoon leader with B/1-27 Infantry Regiment "Wolfhounds," and his platoon were conducting an assessment of the area. (Photo by SGT Brad Willeford)

LTC Richard "Flip" Wilson, of Brooklyn, NY, discusses security concerns Sept. 22 with the Rashidyah Nahia Iraqi Police chief at the nahia police station, northeast of Baghdad. The leaders met to discuss ways Coalition Forces could assist with the problems facing the area. Wilson commands the 1-27 Infantry Regiment "Wolfhounds." (Photo by SGT Brad Willeford)

Children in a Husseniyah marketplace weigh 1LT Joshua Kordela, of Owings Mills, MD, on Sept. 15, while he patrols through this northeast Baghdad town. Kordela is an infantry platoon leader with B/52 Infantry Regiment "Hammer." (Photo by SSG JB Jaso III)

Ambler, PA native, SPC Eugene Dimauro maintains security with a young child on top of a compound in Baghdad's Adhamiyah security district on Sept. 22, while military leaders collect atmospherics below. Dimauro is an infantryman with A/1-27 Infantry Regiment "Wolfhounds." (Photo by SSG JB Jaso III)

Batesville, AR native, SPC Stephen Miller maintains security outside of a Husseniyah school on Sept. 22, ensuring the future leaders of Iraq remain safe from criminal or terrorist activity. Miller is an infantryman with A/1-27 Infantry Regiment "Wolfhounds." (Photo by SSG JB Jaso III)

Children from Baghdad's Adhamiyah security district gather around Fort Ann, NY native, MAJ Patrick Aspland, as he distributes candy on Sept. 23. Aspland is the executive officer for 1-27 Infantry Regiment "Wolfhounds." (Photo by SSG JB Jaso III)

Channahon, IL native, CPL Vern Smith talks with children in the Husseniyah Nahia on Sept. 24, while patrolling through this northeast Baghdad community informing residents of wanted criminals. Smith is a mortarman with A/1-27 Infantry Regiment "Wolfhounds." (Photo by SSG JB Jaso III)

A young child patrols alongside of SPC Christopher Metzler, a native of Sterling, IL, in Istaqlal Qada's Husseniyah town on Sept. 24. Metzler is an infantryman assigned to B/52 Infantry Regiment "Hammer." (Photo by SSG JB Jaso III)

El Paso, TX native, SPC John Ward provides medical assistance to a child that was injured in Istaqlal Qada's Husseniyah town on Sept. 24. Ward is a combat medic with B/52 Infantry Regiment "Hammer." (Photo by SSG JB Jaso III)

RYAN

Army private put Navy dad 'through hell'

By SSG JB Jaso III
1-27 Infantry Regiment "Wolfhounds"

Running away from home, stealing the family's car and taking his father's checkbook are all things of the past. Today, PFC Joseph Biando serves as an infantryman with HHC/1-27 Infantry Regiment "Wolfhounds."

It was a long and bumpy road for Biando to reach where he is today; a successful Soldier serving in his 10th month of a 15-month deployment. He lives and works in the Istaqlal Qada, just northeast of Baghdad.

"I was doing everything bad and nothing good," said the 19-year-old, Chesapeake, VA native, about his childhood. "I put my dad's life through hell! I ran away several times; one time for 18 days."

Biando's father, a U.S. Naval officer with 24 years of service, had equally important issues to face when he wasn't dealing with his son's troublesome nature.

At one point, LT Cmdr. Keith Biando, Joseph's father, deployed in the Persian Gulf on the patrol coastal boat USS Firebolt.

"I was deployed when a lot of stuff was happening with him," said Keith, in regards to his son's free-spirited behavior. "I tried to be a father through a (satellite) phone."

If it wasn't for an argument with his father that resulted in Joseph leaving home for good with no shoes or shirt, he probably wouldn't be serving with distinction in Iraq today. The day after he left his home, he went to the Army recruiting station and enlisted as an infantryman.

"It was the quickest way out of the situation I was in," said Joseph, about joining the Army in June 2007. "I always figured I would go into the military, but never thought the Army."

Still not speaking to his family, Joseph stayed with a friend for two weeks before departing for basic combat training. Even though Keith knew where his son was and what was going on, he decided to let Joseph tell him in his own time.

Finally, when Joseph was in need of clothes and other personal items, he talked to his father about his decision to join the Army, and the entire Biando family met with Joseph on his last night at home before heading to basic training. They didn't talk about the past and the troubles that Joseph had caused; rather they enjoyed their last night together as a Family.

That was the last night they were all together before Joseph would depart for Iraq. Once he completed basic combat training in November 2007, he was assigned to the Wolfhounds at Schofield Barracks, Hawaii. He didn't have much time to enjoy the island as he deployed to Iraq three weeks after his arrival.

"I called my dad, and he told me that it had to be done," Joseph said. "He helped me get situated and told me to be careful."

"I gave him the same guidance I give my young Sailors," Keith said. "Show up on time, and you'll be good!"

Not knowing anyone yet, Joseph boarded the plane in Hawaii to do what he had trained for – combat. "I was scared, nervous, worried, but also excited," said Joseph, about his feelings while flying to Iraq.

Joseph is now 10 months into his deployment the feelings of fear, nervousness and worrisome have all but diminished, and he has embraced his job as a member of the personal security detachment for the Wolfhound commander.

It's no easy task standing in 120-degree summer temperatures wearing the required protective gear, while others get to be inside air-conditioned buildings.

However, Joseph realizes the importance of his duties.

"It's an important task," he said about his job. "We secure the battalion's leadership in order to allow them to rebuild and reconstruct Iraq."

Because of the security he provides, the Wolfhounds have been able to assist the Government of Iraq on the road to reconstruction. If it's a difficult journey down that road – Joseph can relate.

Joseph was selected as the Wolfhound Soldier of the month in July and received the Army Achievement Medal for meritorious service in Iraq.

"He was so proud to be selected," said his father. "I'm proud of him."

"It felt good," said Joseph, after receiving the award. "I felt special for about 20 minutes until I went on mission – then it was back to work."

Joseph has another mission – to make his dad proud.

"I just want to make my dad proud," Joseph said. "I want to redeem myself, but it's going to take awhile. I caused my dad a lot of stress."

But Keith said he has no resentment over his son's troubled youth.

"I never had any resentment or ill feelings for what he did," the lieutenant commander said of his son's childhood actions. "A dad can't have any. I just tried to steer him on the right path and away from the bad.

"He's always trying to make me proud, and I recognize that," Keith said. "I recognize his maturity and I'm so proud of him."

Now that he knows his father is proud of him, Joseph wants to continue to be the best he can and try to live up to his father.

"I apologize for my childhood, and I'm proud of my dad," Joseph said.

"I wish I could accomplish what he has accomplished in his 24 years in the Navy."

Even though Joseph has several years until he could accomplish all that his dad has, he is defiantly on the road to success, looking at his childhood nightmares in the rearview mirror.

PFC Joseph Biando, a Chesapeake, VA native, is happy to see the reaction of children in the Taji Qada, northwest of Baghdad Apr. 14, as their school reopens after being reconstructed by the Government of Iraq's Ministry of Education. Biando is a mortarman assigned to HHC/1-27 Infantry Regiment "Wolfhounds." (Photo by SSG JB Jaso III)

(Left) 1LT Tommy Ryan, of Seminole, FL, ensures that CPL Jerimiah Sullivan, of St. Joseph, MI is not injured following an improvised explosive device that struck their Stryker armored vehicle on Sept. 26 in the Husseniyah town of the Istaqlal Qada. Ryan is an infantry platoon leader and Sullivan is a vehicle commander with A/1-27 Infantry Regiment "Wolfhounds." (Photo by SSG JB Jaso III)

(Right) 1LT Tommy Ryan, of Seminole, FL, ensures that the Soldiers riding in the back of his Stryker armored vehicle are not injured following an improvised explosive device that struck their vehicle in Istaqlal Qada's Husseniyah town on Sept. 26. Ryan is an infantry platoon leader with A/1-27 Infantry Regiment "Wolfhounds." (Photo by SSG JB Jaso III)

(Left) 1LT Tommy Ryan, of Seminole, FL, prepares to exit his Stryker armored vehicle following an improvised explosive device that struck his vehicle on Sept. 26 in the Husseniyah town of the Istaqlal Qada. Ryan is an infantry platoon leader with A/1-27 Infantry Regiment "Wolfhounds." (Photo by SSG JB Jaso III)

Carl Junction, MO native, SPC David Mensink talks with a sheik in Baghdad's Adhamiyah security district about improvements that are being made in this northeast Baghdad community on Sept. 26. Mensink is an infantryman with A/1-27 Infantry Regiment "Wolfhounds." (Photo by SSG JB Jaso III)

San Bernardino, CA native, SGT Jared Bruyneel waves to children as he passes them during a patrol to identify generators provided by terrorist groups in Istaqlal Qada's Husseniyah town on Sept. 28. Bruyneel is an infantry team leader with A/1-27 Infantry Regiment "Wolfhounds." (Photo by SSG JB Jaso III)

Sasakwa, OK native, SPC Arthur Smith maintains security in an industrial district in Baghdad's Adhamiyah security district on Sept. 26, ensuring the safety of residents. Smith is an infantryman with A/1-27 Infantry Regiment "Wolfhounds." (Photo by SSG JB Jaso III)

Greenville, SC native, SSG Rodriguez Cruell collects information from residents about generators that are provided to the community by terrorist groups on Sept. 28. Cruell is an infantry squad leader with A/1-27 Infantry Regiment "Wolfhounds." (Photo by SSG JB Jaso III)

Kenosha, WI native, SPC Climmie Gaston reacts to small arms fire that is directed at his platoon's location in Istaqlal Qada's Husseniyah town on Sept. 28. Gaston is an infantryman with A/1-27 Infantry Regiment "Wolfhounds." (Photo by SSG JB Jaso III)

Seminole, FL native, 1LT Tommy Ryan ensures that his men are okay and reacting properly to small arms fire directed at his platoon in the Husseniyah town, while a Stryker armored vehicle approaches to assist on Sept. 28. Ryan is an infantry platoon leader with A/1-27 Infantry Regiment "Wolfhounds." (Photo by SSG JB Jaso III)

Valparaiso, IN native, PFC Christopher Mola patrols through the Husseniyah town on Sept. 28, identifying generators that are provided to the community by terrorist organizations. Mola is an infantryman with A/1-27 Infantry Regiment "Wolfhounds." (Photo by SSG JB Jaso III)

Chicago native, 1LT Aaron Kletzing; and Mililani, HI native, CPT Peter Walther, review the map of a Husseniyah town marketplace on Sept. 29. Kletzing is the fire support officer, while Walther is the executive officer for A/1-27 Infantry Regiment "Wolfhounds." (Photo by SSG JB Jaso III)

Kennesaw, GA, native, SPC John Pearson plays a video game at an arcade in Istaqlal Qada's Husseniyah town on Sept. 29, while children look on with excitement. Pearson is an assistant team leader with the 1620th Tactical Psychological Operations Detachment, 312th Tactical Psychological Operations Company. (Photo by SSG JB Jaso III)

SGT Michael Robinson, of Lake Forest, CA, pretends to surf under the crossing sabers at Baghdad's International Zone on Sept. 29. Robinson is a sniper team leader with C/1-27 Infantry Regiment "Wolfhounds." (Photo by SGT Karl Schell)

Winchendon, MA native, SPC Garrett Simoneau patrols through the streets of the Husseniyah Nahia on Sept. 30, ensuring the safety and security of residents in this northeast Baghdad region. Simoneau is an infantryman with A/1-27 Infantry Regiment "Wolfhounds." (Photo by SSG JB Jaso III)

Sasakwa, OK native, SPC Arthur Smith patrols through the Husseniyah town of the Istaqlal Qada on Sept. 30, ensuring the safety of residents in this northeast Baghdad community. Smith is an infantryman with A/1-27 Infantry Regiment "Wolfhounds." (Photo by SSG JB Jaso III)

SFC Ernest Oberst

Wolfhound leader is part of team

PFC Lyndsey Dransfield
SBCT PAO, 25th Inf. Div., MND-B

During a World War II battle in Holtzwihr, France, Audie Murphy, a wounded U.S. Soldier, climbed inside of a burning tank, took a spot behind a .50-caliber machine gun and fired until the enemy was vanquished.

At the age of 19, he was awarded the Medal of Honor and, to this day, is an inspiration to military leaders, Soldiers – and SFC Ernest Oberst.

"As a kid, I used to watch the Audie Murphy Biography 'To Hell and Back' on TV, and I wanted to be a Soldier," said Oberst, native of Gladstone, MI, who joined the Army three months after graduating high school. "He was my hero."

Oberst is a platoon sergeant with 1st Platoon, A/1-27 Infantry Regiment "Wolfhounds."

He is responsible for all 33 Soldiers in his platoon. He's required to take care of administration paperwork, lead patrols, make sure each Soldier has the equipment they need and take care of any issues that may arise, whether personal or work related. "I do anything a mom or dad does," he said jokingly.

Oberst has served in the U.S. Army for 14 years and developed a great deal of knowledge. He said he believes the best way to lead his Soldiers is by setting an example for them and sharing the load. "You can't just supervise, you have to be a part of the team," he said "If my Soldiers are out digging for caches, you won't find me sitting in the truck; I'm an able body that can work, and I'll be out there digging with them."

His Soldiers describe him as a well-rounded leader who maintains discipline and has the ability of doing the right thing at the right time, every time.

"No matter what the mission is, Oberst is the first one on the ground and the last one back in the vehicle. That's just his style," said CPL Zachary Manuel, a Pickerington, OH native, with A/1-27 Inf. Regt.

SGT Lucas Collins, native of Manchester, GA, said Oberst will give any Soldier a chance to succeed. "When I came into this unit, I had two blown-out knees," said Collins, a team leader. "In the infantry, you are looked at as done. I was going to be chaptered out, but he gave me the chance to get better."

Three years later, Collins is awaiting promotion to staff sergeant and said a great amount of what he's learned is attributable to Oberst.

"He has shown me that taking care of my Soldiers is number one," he said. "And if something needs to be done, make sure it gets done."

Oberst not only makes sure the job gets done, but he ensures the job gets done right the first time.

Throughout the years, Oberst has gained the respect of his fellow Soldiers and those he mentors as well as discovered the key elements essential to be good leader. "He requires the best out of his men and expects nothing less," said Manuel. "For that reason, he gets the best."

TOYOTA
الله أكبر

SPC Charles Joseph, of Kerrville, TX, stand with children near their new playground in Hor al Bosh, a village northwest of Baghdad Oct. 1. Soldiers from B/52 Infantry Regiment conducted a joint patrol of the Taji Market area with Iraqi Police, Oct 1. During the patrol, they checked progress of road paving, assessed water pumps, and donated playground equipment. Joseph is a Radio Telephone Operator assigned to B/52 Inf. Regt. (Photo by PFC James Childs)

Soldiers from B/52 Infantry Regiment train seven Iraqi Police officers on vehicle searches at the recently opened Hammamiat IP Station near Hor al Bosh, a village northwest of Baghdad Oct. 1. Topics of training included checkpoint security, search procedures and leader responsibilities. This training was only one stop in the day's patrol. (Photo by SPC Charles Joseph)

SSG Brandon Tulloch, of Marriotsville, MD, watched an Iraqi Police officer demonstrate the clearing of the M4 rifle while his fellow officers observe in Hor al Bash, northwest of Baghdad, Sept. 24. Tulloch is a squad leader assigned to B/52 Infantry Regiment. (Photo by SPC Charles Joseph)

PFC Edgar Gutierrez, of Santa Rosa, CA, touches a rose at a school in Baghdad's Adhamiyah security district on Oct. 1. Gutierrez is a mortarman assigned to HHC/1-27 Infantry Regiment "Wolfhounds." (Photo by SSG JB Jaso III)

SPC Christopher Metzler, of Sterling, IL, patrols through a busy recreation center in the Husseniyah town of the Istaqlal Qada on Oct. 3. Metzler is an infantryman assigned to B/52 Infantry Regiment "Hammer." (Photo by SSG JB Jaso III)

1LT Joshua Kordela, of Owings Mill, MD, conducts a pre-mission brief on Oct. 3, informing his soldiers of their mission and intent. Kordela is an infantry platoon leader assigned to B/52 Infantry Regiment "Hammer." (Photo by SSG JB Jaso III)

CPT Peter Walther, of Mililani, HI, lines up his shot before sinking the six ball in the corner pocket during a patrol in the Istaqlal Qada on Oct. 3, northeast of Baghdad. Walther is the executive officer for A/1-27 Infantry Regiment "Wolfhounds." (Photo by SGT Brad Willeford)

(Left) SGT Andrew Lambright, of Eugene, OR, is ready to enjoy a very large Rice Krispies Treat at Joint Security Station Istaqlal on Oct. 4. Lambright is an infantry team leader with A/1-27 Infantry "Wolfhounds." (Photo by PFC Stephen Miller)

(Right) Fort Ann, NY native MAJ Patrick Aspland maintains his golfing skills by playing on a video game at a Joint Security Station in the Istaqlal Qada on Oct. 6. Aspland is the executive officer for the 1-27 Infantry Regiment "Wolfhounds." (Photo by SSG JB Jaso III)

(Left) Miami native, SSG Donald Fabian prepares to drive a golf ball down the fairway at a Joint Security Station in the Istaqlal Qada on Oct. 6. Fabian is a team leader with the 1620th Tactical Psychological Operations Detachment, 312th Tactical Psychological Operations Company, Multi-National Division-Baghdad. (Photo by SSG JB Jaso III)

Tomball, TX native, SPC Timothy Smith patrols through a marketplace in the Subak Sur Nahia on Oct. 7, allowing residents to live their lifes without fear of criminal or terrorist activity. Smith is an infantryman assigned to B/52 Infantry Regiment "Hammer." (Photo by SSG JB Jaso III)

SGT Tony Macias, of Wauchula, FL, shows children, in the Husseniyah Nahia, his age after being asked while patrolling this community, Oct. 7. Macias is an infantry squad leader with A/1-27 Infantry Regiment "Wolfhounds." (Photo by SSG JB Jaso III)

SGT Michael Rios, of Sebastian, TX, says hello to children in the Husseniyah Nahia, Oct. 7, while conducting an early evening patrol ensuring the safety of these future leaders of Iraq. Rios is an infantry squad leader with A/1-27 Infantry Regiment "Wolfhounds."
(Photo by SSG JB Jaso III)

Honolulu native, SSG Kalei Akana maintains security while patrolling through the Subak Sur Nahia on Oct. 7, gathering information on criminals in this northern Baghdad community. Akana is an infantry squad leader assigned to B/52 Infantry Regiment "Hammer." (Photo by SSG JB Jaso III)

Soldier in Focus

By SFC Christina Bhatti
2nd SBCT PAO, 25th Inf. Div., MND-B

A company commander's success is based on the accomplishments of their Soldiers. For CPT Brian Sweigart, a native of San Antonio, and commander of HHC/1-27 Infantry Regiment "Wolfhounds," that is an idea he takes seriously. "I take personal pride in what my Soldiers do," he said. "My job is to talk to them and mentor them. It's great to know that I have a hand in shaping their successes. If they are successful that means I am successful."

Sweigart received his commission eight years ago after graduating from the United States Military Academy at West Point in West Point, NY. He followed in the footsteps of his father and grandfather who also served in the military.

"It is something I always wanted to do," he said. "I saw them serve proudly and that is a tradition I wanted to continue."

He deployed in support of Operation Iraqi Freedom with the 2nd SBCT in December of 2007. Since the unit's arrival he said he has seen many positive accomplishments from his unit, but also the Iraqi people.
"The Iraqi people have now begun to take responsibility and accountability for themselves," he said.

Sweigart said he regularly sees Iraqi inspectors checking on projects instead of U.S. commanders "without engineering degrees inspecting with only a checklist from the unit's engineers." This is a turning point which he said they can be proud of, and hopes these kinds of steps forward will continue for the country of Iraq and is proud he is a part of the progress.

CPT Brian Sweigart, of San Antonio, TX, commands the Headquarters and Headquarters Company, 1st Battalion, 27th Infantry Regiment "Wolfhounds," 2nd Stryker Brigade Combat Team "Warrior," 25th Infantry Division, Multi-National Division-Baghdad. (Photo by SFC Christina Bhatti)

Hillsboro, TN, native, SGT Jamie Norris patrols through the Subak Sur Nahia on Oct. 7, ensuring the safety of residents in this northern Baghdad community. Norris is an infantry team leader with A/1-27 Infantry Regiment "Wolfhounds." (Photo by SSG JB Jaso III)

Sebastian, TX native, SGT Michael Rios patrols through a neighborhood in the Husseniyah Nahia on Oct. 7, attempting to locate the home of a high-value individual. Rios is an infantry team leader with A/1-27 "Wolfhounds." (Photo by SSG JB Jaso III)

Peachtree, GA native, SPC Collin Kelly maintains security during an early evening patrol through the Husseniyah Nahia on Oct. 7. Kelly is an infantryman with A/1-27 Infantry Regiment "Wolfhounds." (Photo by SSG JB Jaso III)

Seminole, FL native, 1LT Tommy Ryan talks with the unit's bilingual bicultural advisor on Oct. 7, after inspecting solar lights that were funded and installed by the GoI. Ryan is an infantry platoon leader with A/1-27 "Wolfhounds." (Photo by SSG JB Jaso III)

Circleville, OH native, SGT Timothy Miller patrols past an Iraqi political party headquarters in the Husseniyah town on Oct. 9. Miller is an infantry team leader with A/1-27 Infantry Regiment "Wolfhounds." (Photo by SSG JB Jaso III)

Boca Raton, FL native, SSG Christopher Bolwell patrols down a Husseniyah town street on Oct. 9, disseminating Baghdad Now newspapers. Bolwell is an infantry squad leader with A/1-27 Infantry Regiment "Wolfhounds." (Photo by SSG JB Jaso III)

SPC Donald Lane, of Kansas City, MO, maintains security while soldiers search a house suspected of hiding illegal weapons on Oct. 9 in Baghdad's Adhamiyah security district. Lane is an infantryman with A/1-27 "Wolfhounds." (Photo by SSG JB Jaso III)

SPC Eric Waddle, of Bedford, IN, overlooks the Tigris River while military leaders and water experts evaluate a Compact Water Treatment Unit in the Istaqlal Qada, Oct. 9. Waddle is a mortarman assigned to HHC/1-27 Infantry Regiment "Wolfhounds." (Photo by SSG JB Jaso III)

Batesville, AR native, SPC Stephen Miller maintains security in the Husseniyah Nahia on Oct. 16, while children climb a gate. Miller is an infantryman with A/1-27 Infantry Regiment "Wolfhounds." (Photo by SSG JB Jaso III)

CPT Lucas Yoho, a Long Beach, CA native, talks to residents about the importance of a rubbish removal project that is occurring in Istaqlal Qada's Husseniyah town on Oct. 16. Yoho commands A/1-27 Infantry Regiment "Wolfhounds." (Photo by SSG JB Jaso III)

CPT Lucas Yoho, a Long Beach, CA native, talks about improvements made to Istaqlal Qada's Husseniyah town on Oct. 16. Yoho commands A/1-27 Infantry Regiment "Wolfhounds." (Photo by SSG JB Jaso III)

Calverton, NY native, PFC Kyle Lauth patrols through a Husseniyah Nahia neighborhood on Oct. 16 during Operation Gunslinger Bonzai XVI. Lauth is an infantryman with A/1-27 Infantry Regiment "Wolfhounds." (Photo by SSG JB Jaso III)

SSG Rudy Mello, a San Antonio, native, looks at a newspaper with a man Oct. 19 in the Rashidyah Nahia, northeast of Baghdad, during a patrol through a market located northeast of Baghdad. Mello serves as an infantry squad leader with B/1-27 Infantry Regiment "Wolfhounds." (Photo by SGT Brad Willeford)

Wolfhounds, GoI work to rebuild education

By SSG JB Jaso III
1st Bn., 27th Inf. Regt., 2nd SBCT, 25th Inf. Div., MND-B

Leaders from Multi-National Division – Baghdad's 1-27 Infantry Regiment "Wolfhounds," joined officials from the Government of Iraq's Ministry of Education, community leaders, Iraqi National Police and more than 500 students at the Bilal School in Baghdad's Adhamiyah security district, Oct. 11.

The large crowd gathered to celebrate the opening of the area's largest school, which recently received a much needed facelift.

"Just three months ago, the school was uninhabitable; it was in disarray," said 1LT John Busuego, an Ontario, CA, native, and the Wolfhounds essential services coordinator. "The school was not in use for a long time and the Ministry of Education wanted to refurbish it for the area."

Construction workers braved the summer heat and completed the nearly $200,000 worth of projects in only three months.

The construction workers did all the necessary repairs, inside and out, Busuego said. They installed a new main gate, new lighting, electrical, a sewer system and painted.

"Definitely, one of the best looking schools I've seen here in Iraq," said LT Timothy Schul, a California, MD, native. Schul is the Wolfhounds Civil Affairs officer who observed progress and knows the importance of this school to the area.

"The GoI has gained confidence, and they are concerned with education and everyone was pleased to see the cooperation, which is a step in the right direction," he said.

The step in the right direction began with the ribbon cutting ceremony, symbolizing the official opening of the school.

Iraqi Col. Ahmed Mohammed Mohammed, commander of the 2nd Brigade, 1st Iraqi National Police Division, cut the ribbon and all present were eager to enter the school too see the progress made.

After walking through the newly opened school, community and government leaders spoke of the importance of the school's opening. The police distributed Iraqi flag stickers, flags, soccer balls and soccer jerseys to the students.

"The students all stuck the Iraqi flag stickers on their chest," said 1LT Aaron Kletzing, a Chicago native, and A Co.'s fire support officer. "It was so surreal."

As the festivities concluded, Kletzing continued to watch the residents from the area approaching the city council members.

"People were coming up to the city council members saying that it was a good job," he said. "The people were thanking the council for taking care of them."

"The area hasn't had such a fulfilling happy experience in some time," Kletzing said. "There are probably five things that I will always remember about this deployment," Kletzing said. "This is one of them; I was really impressed."

Students from the Bilal School hold Iraqi flags distributed by the 1st Brigade, 2nd Iraqi National Police Division, before the grand opening of the school in Baghdad's Adhamiyah security district, Oct. 11. (Photo by SSG JB Jaso III)

Valparaiso, IN native, PFC Christopher Mola maintains security outside of a government building during a city council meeting in the Husseniyah Nahia on Oct. 20. Mola is an infantryman with A/1-27 Infantry Regiment "Wolfhounds." (Photo by SSG JB Jaso III)

Stanley, VA native, SPC Timothy Lewis carries Salina, KS native SPC Mackenzie Diaz; and Lake Forest, CA native, SGT Michael Robinson, to their Stryker armored vehicle prior to a patrol on Oct. 21. Lewis, Diaz and Robinson are infantrymen with C/1-27 Infantry Regiment "Wolfhounds." (Photo by SGT Karl Schell)

The CRT with a newly-fielded Stryker Recovery System on Oct. 24. The 2/25 Stryker Brigade Combat Team was the first unit in the Army to be fielded this new recovery system specifically designed to tow and recover disabled Stryker armored vehicles and other large military vehicles. (Courtesy Photo)

SSG George Canfil, of La Mesa, CA, talks to a Subak Sur Nahia resident about the importance of notifying coalition forces of the whereabouts of criminals and terrorists. Canfil is an intelligence team leader with A/1-27 "Wolfhounds."(Photo by SSG JB Jaso III)

SFC Jesse Wyant, of Geneva, NY, who serves as a platoon sergeant with C/1-27 Infantry Regiment "Wolfhounds", greets students at the al Hayal School in Bassam, northwest of Baghdad, Oct. 26. Soldiers from C Company distributed school supplies to the approximately 400 students and 14 teachers. (Photo by SFC Christina Bhatti)

Soldiers from 1-27 Infantry Regiment "Wolfhounds," enjoy the "Comics Ready to Entertain" tour, which stopped at Joint Security Station Istaqlal on Oct. 20. (Photo by SSG JB Jaso III)

Comedian, Bengt Washburn entertains soldiers from 1-27 Infantry Regiment "Wolfhounds" at Joint Security Station Istaqlal on Oct. 20, during the "Comics Ready to Entertain" tour.
(Photo by SSG JB Jaso III)

Soldiers from 2nd Platoon, A/1-27 Infantry Regiment "Wolfhounds," continue their tradition of celebrating Halloween on Oct. 31, no matter where they are - Hawaii or Iraq. (Photo by SSG JB Jaso III)

A Soldier from B/1-27 Infantry Regiment "Wolfhounds", stands guard on the roof of the Fahama Primary School in Istaqlal, northeast of Baghdad, during an opening celebration Oct. 23. Approximately $200,000 was spent on structural repairs, rewiring, windows, doors and painting. (Photo by SFC Christina Bhatti)

SGT Eli Bojorquez, of Lincoln, OR, a personal security detachment noncommissioned officer assigned to HHC/1-27 Infantry Regiment "Wolfhounds", watches over a group of children at the Fahama Primary School in Istaqlal, northeast of Baghdad, during an opening celebration Oct. 23. (Photo by SFC Christina Bhatti)

SPC Michael Waldie, of Gualala, CA, assigned to B/1-27 Infantry Regiment "Wolfhounds", hands a backpack full of supplies to a student after the Fahama Primary School opening celebration Oct. 23. The school in the Istaqlal Qada, northeast of Baghdad, underwent approximately $200,000 worth of renovations. (Photo by SFC Christina Bhatti)

SPC Brian Cochran, of Dallas, TX, who serves as an infantryman with C/1-27 Infantry Regiment "Wolfhounds", helps unload a trailer full of donated school supplies in Basam, northwest of Baghdad, Oct. 26. The supplies were donated by various organizations in the United States.
(Photo by SFC Christina Bhatti)

PFC Brandon Pillen, of Olmstead, OH, who serves as an infantryman with C/1-27 Infantry Regiment "Wolfhounds", hands out school supplies at the al Hayal School in Bassam, northwest of Baghdad, Oct. 26. The supplies were donated by various organizations in the United States.
(Photo by SFC Christina Bhatti)

Soldiers from 2nd Platoon, A/1-27 Infantry Regiment "Wolfhounds" celebrate their Halloween in Iraq by painting their faces on Oct. 31. (Photo by SSG JB Jaso III)

Brooklyn, NY native, SPC Tyrell McCurbin receives a Halloween goodie bag that was sent by the unit's family readiness group on Oct. 31. McCurbin is a intelligence analyst for the 1-27 Infantry Regiment "Wolfhounds." (Photo by 1LT Joseph Mckinney)

Utilizing a fly swatter as a sword; a cut balaclava as a mask; and a shirt as a cape, Midlothian, TX native, PFC Richard Plasket dresses as "Zorro" for Halloween on Oct. 31 at Joint Security Station Istaqlal. Plasket is an infantryman with A/1-27 Infantry Regiment "Wolfhounds." (Photo by SSG JB Jaso III)

Gimlets help plan to rebuild one of Aqur Quf's oldest historical landmarks

By SGT Karl Schell
1st Bn. 21st Inf. Regt., 2nd SBCT, 25th Inf. Div., MND-B

Plans are in the works to help restore one of Aqur Quf's oldest landmark: the Ziggurat.

The Ziggurat of Aqur Quf, which lies west of Baghdad, is one of the best preserved religious landmarks of its kind in Iraq. It dates back to 1400 B.C. and was built as a monument to the local religion.

Soldiers from C/1-27 Infantry Regiment, attached to 1st Battalion, 21st Infantry Regiment, are working in conjunction with the citizens of Iraq and branches of the Government of Iraq to rebuild and repair the site and buildings in an effort to reopen the Ziggurat as a tourist attraction.

Once open, it will offer a promising future of commerce from around the Middle East. The rebuilding alone will offer work for the Iraqi citizens through the rebuilding, and once the rebuilding is complete and open to the public, it will provide the economy with a steady working environment and a constant flow of tourists, which will provide business to other merchants in Aqur Quf, said CPT Devon McRainey, native of Dublin,OH, and C Co. executive officer.

"It feels good to be a part rebuilding important historical sights of Iraq's culture," he said.

The project will begin with the repair to the fence that surrounds the Ziggurat. The next step is to ensure the safety of the historical artifacts, which will be brought back to the site from a holding place in Baghdad. The artifacts are currently held by the Ministry of Ruins until the completion of the repairs.

Once the project is finished the Aqur Quf Ziggurat will be the home to a new restaurant, park and tours guided by historians.

"This will stand as an example that Iraqi security is improving," said CPT David Uthlaut, a native of Charleston, SC, and C Co. commander. "If we leave any mark on the people of Iraq, this will be a great one."

Soldiers from C/1-27 Infantry Regiment "Wolfhounds," attached to 1st Battalion, 21st Infantry Regiment, walk up the stairs with members of the Aqur Quf Historical Society to view the Aqur Quf Ziggurat, west of Baghdad, Oct. 14. (Photo by SGT Karl Schell)

The Aqur Quf Ziggurat, west of Baghdad, is considered one of the most well preserved religious historical monuments in Iraq. Soldiers from C/1-27 Infantry Regiment "Wolfhounds" are assisting the Aqur Quf Historical in repairing the grounds and building surrounding the Ziggurat in order to be able to open the site to tourism. (Photo by SGT Karl Schell)

San Diego native, SPC Eric Gonzalez maintains security outside of a city council meeting in Baghdad's Adhamiyah security district on Nov. 1, allowing Government of Iraq officials to continue improving the quality of life for residents in this northern Baghdad community. Gonzalez is a communications specialist with A/1-27 Infantry Regiment "Wolfhounds."
(Photo by SSG JB Jaso III)

SGT Antwan Scales, of Bellwood, IL, speaks to a young man about a school project that is pending in a Sab al Bor Nahia neighborhood, located northwest of Baghdad on Nov. 1. Scales is an infantry team leader assigned to C/1-27 Infantry Regiment "Wolfhounds."
(Photo by SPC John Ahn)

Moore, OK native, SSG Michael McMurray ensures that proper security measures are emplaced at a Sab al Bor Nahia primary school, located northwest of Baghdad on Nov. 1. McMurray is an infantry squad leader assigned to C/1-27 Infantry Regiment "Wolfhounds."
(Courtesy Photo)

Hamilton, OH native, CPT Chad Chenoweth gets ready to send in his ballot on Camp Taji on Nov. 1. Chenoweth works as the logistics officer for the 1-27 Infantry Regiment "Wolfhounds."
(Courtesy Photo)

CSM William Hain; MSG Timothy Foster; 1SG Timothy Jackson; and SGT Joshua Oden were the backbone to JSS Istaqlal's expansion, allowing the "Wolfhound" soldiers to live more comfortable and healthy. (Photo by CPT Chad Chenoweth)

1SG Timothy Jackson, of Dryden, NY, ensures that new containerized housing units (CHUs) are properly emplaced at JSS Istaqlal, Nov. 3, ensuring the continued improvements of living conditions. Jackson is the first sergeant for HHC/1-27 Infantry Regiment "Wolfhounds." (Photo by CPT Chad Chenoweth)

Onxard, CA native, SSG Felipe Torres takes a break after helping to put in living trailers on the Joint Security Station in the Istaqlal Qada on Nov. 3. Torres serves as a mortar section leader with HHC/1-27 Infantry Regiment "Wolfhounds." (Photo by CPT Chad Chenoweth)

Poulsboro, WA native, SPC Charles Vanderhalf stands guard outside a grocery store in Baghdad's Adhamiyah security district on Nov. 4. Vanderhalf is a mortarman with A/1-27 Infantry Regiment "Wolfhounds." (Photo by SSG JB Jaso III)

Irwin, PA native, PFC Corey Show, maintains security while patrolling through a business district in the Husseniyah town on Nov. 4. Show is an infantryman with A/1-27 Infantry Regiment "Wolfhounds." (Photo by SSG JB Jaso III)

Sacramento, CA native, SGT Steve Thompson patrols the Sab al Bor Nahia countryside on Nov. 5, ensuring the continued security in this northwest Baghdad region. Thompson is an infantry team leader assigned to C/1-27 Infantry Regiment "Wolfhounds." (Photo by SPC John Ahn)

Owings Mills, MD native, 1LT Joshua Kordela asks a manikin to dance while patrolling through the Husseniyah Nahia on Nov. 4. Kordela is an infantry platoon leader with B/52 Infantry Regiment, attached to A/1-27 Infantry Regiment "Wolfhounds." (Photo by SSG JB Jaso III)

Brooklyn, NY native, SSG Harry Jenkins; and Saint Clair Shores, MI native, SSG Joseph Stalinski, divide sectors of responsibility during a civic meeting being held in the Sab al Bor Nahia, located northwest of Baghdad, on Nov. 5. Jenkins and Stalinski are both infantry squad leaders with C/1-27 Infantry Regiment "Wolfhounds." (Photo by SPC John Ahn)

Jasper, GA native, PFC Benjamin Garcia says hello to children in the Sab al Bor Nahia on Nov. 5, while conducting a morning patrol ensuring the safety of this northwest Baghdad region. Garcia is an infantry automatic rifleman with C/1-27 Infantry Regiment "Wolfhounds." (Photo by SPC John Ahn)

Lewiston, ID native, PFC Chance Shrader keeps an eye out for trouble during a civic meeting held during a beautiful fall morning in the Sab al Bor Nahia, located northwest of Baghdad, on Nov. 5. Shrader is an infantry automatic rifleman assigned to C/1-27 Infantry Regiment "Wolfhounds." (Photo by SPC John Ahn)

Saint Clair Shores, MI native, SSG Joseph Stalinski patrols the wide spread neighborhoods of the Sab al Bor Nahia, located northwest of Baghdad, on Nov. 5. Stalinski is an infantry squad leader with C/1-27 Infantry Regiment "Wolfhounds." (Photo by SPC John Ahn)

SPC Jeremy Picnic, native of Fairview, NJ, takes point in order to guide his platoon on a dismounted patrol through the Sab al Bor Nahia, located northwest of Baghdad, on Nov. 5. Picnic is an infantryman assigned to C/1-27 Infantry Regiment "Wolfhounds." (Photo by SPC John Ahn)

SSG Harry Jenkins, of Brooklyn, NY, gives SPC David Lopez, of San Antonio, a sector of fire as they maintain security during a halt while conducting a dismounted patrol in the Sab al Bor Nahia, located northwest of Baghdad, on Nov. 5. Jenkins is an infantry squad leader with C/1-27 Infantry Regiment "Wolfhounds." (Photo by SPC John Ahn)

Dallas native, SPC Brian Cochran guards an entrance to a civic meeting on November 5 in the Sab al Bor Nahia, located northwest of Baghdad. Cochran is an infantryman with C/1-27 Infantry Regiment "Wolfhounds." (Photo by SPC John Ahn)

Cranwell, CT native, SPC Robert Haswell stops to listen and look for any potential threats during a dismounted patrol through a Sab al Bor Nahia neighborhood, located northwest of Baghdad, on Nov. 5. Haswell is an infantryman assigned to C/1-27 Infantry Regiment "Wolfhounds." (Photo by SPC John Ahn)

Hillsboro, TN native, SGT Jamie Norris maintains security outside a Husseniyah Nahia home, alongside an Iraqi National Police officer on Nov 6, during Operation Gunslinger Bonzai XX. Norris is an infantry team leader with A/1-27 Infantry Regiment "Wolfhounds." (Photo by SSG JB Jaso III)

Jasper, GA native, PFC Benjamin Garcia takes a knee around a wall in order to maintain security during a halt while conducting a dismounted patrol in the Sab al Bor Nahia, located northwest of Baghdad on Nov. 5. Garcia is an infantry automatic rifleman with C/1-27 Infantry Regiment "Wolfhounds." (Photo by SPC John Ahn)

Long Beach, CA native, CPT Lucas Yoho finalizes the plan for Operation Gunslinger Bonzai XX with a leader from the 2nd Brigade, 1st Iraqi National Police Division on Nov. 6 in the Husseniyah Nahia. Yoho commands A/1-27 Infantry Regiment "Wolfhounds." (Photo by SSG JB Jaso III)

PFC Michael Dunkel, a Waukegan, IL native, maintains security while members of the 2nd Brigade, 1st Iraqi National Police Division inspect Husseniyah Nahia homes for illegal weapon caches during Operation Gunslinger Bonzai XX on Nov. 6. Dunkel is an infantryman with A/1-27 Infantry Regiment "Wolfhounds." (Photo by SSG JB Jaso III)

San Antonio native, SPC David Lopez patrols the streets of the Sab al Bor Nahia during a dismounted patrol on Nov. 5. Lopez is an infantryman assigned to C/1-27 Infantry Regiment "Wolfhounds." (Photo by SPC John Ahn)

Children from the Husseniyah Nahia talk with Riverhead, NY native SPC Michael Ortiz during Operation Gunslinger Bonzai XX, on Nov. 6. Ortiz is an infantryman with A/1-27 Infantry Regiment "Wolfhounds." (Photo by SSG JB Jaso III)

PFC Michael Dunkel, a Waukegan, IL native, patrols past sheep during Operation Gunslinger Bonzai XX on Nov. 6, a mission focused at identifying and removing illegal weapons from the Husseniyah Nahia. Dunkel is an infantryman with A/1-27 Infantry Regiment "Wolfhounds." (Photo by SSG JB Jaso III)

Pensacola, FL native, CPT Nicholas Pines speaks to a leader about an ongoing canal cleaning project in the Sab al Bor Nahia, on Nov.6. Pines is an infantry platoon leader assigned to C/1-27 Infantry Regiment "Wolfhounds." (Photo by SPC John Ahn)

During Operation Gunslinger Bonzai XX, Riverhead, NY, native, SPC Michael Ortiz notices a young Husseniyah Nahia child wearing a New York Giants sweatshirt on Nov. 6. Ortiz is an infantryman with A/1-27 Infantry Regiment "Wolfhounds." (Photo by SSG JB Jaso III)

(Left) SGT Antwan Scales, of Bellwood, IL, maintains security during a canal cleaning project inspection on Nov. 6 in the Sab al Bor Nahia, allowing water irrigation to improve in this northwest Baghdad community. Scales is an infantry team leader with C/1-27 Infantry Regiment "Wolfhounds." (Photo by SPC John Ahn)

(Right) SPC Seth Wood, of McKinney, TX, dismounts to inspect the area for improvised explosive devices in the Sab al Bor Nahia on Nov. 6. Wood is an infantryman with A/1-27 Infantry Regiment "Wolfhounds." (Photo by SPC John Ahn)

(Left) SGT Antwan Scales, of Bellwood, IL, maintains security at a Sons of Iraq checkpoint during a manning inspection on Nov. 6. Scales is an infantry team leader with C/1-27 Infantry Regiment "Wolfhounds." (Photo by SPC John Ahn)

2 MND-B Soldiers are truly 'Brothers in Arms'

By SPC Curtis Riffle
B/52 Infantry Regiment, 2nd SBCT, 25th Inf. Div., MND-B

CAMP TAJI, Iraq - Serving in the Army can be a long, difficult road. For many, the best way to survive is to embrace friends and coworkers along the way. Fellow Soldiers often become brothers. For two Multi-National Division – Baghdad Soldiers however, they are brothers in the literal sense.

Such is the case for SGT Donnell Macklin and CPT Lionel Macklin, who were happily reunited Nov. 7 at Camp Taji to commemorate Donnell's commitment to remain a part of the Army Team.

Donnell serves as a forward signal support noncommissioned officer with B/52 Infantry Regiment "Hammer" which is based out of Hawaii. Lionel, his older brother, serves as a platoon leader with the 4th Infantry Division's A/3-4 Aviation Squadron, Combat Aviation Brigade, which is based out of Fort Hood, TX.

The two are natives of Ozark, AL, and are intimately aware of the bonds that bring "Brothers in Arms" together. For the two of them, they were able to take advantage of the unique opportunity for the older brother to officially reenlist his younger brother as he renews his commitment to his other Brothers in Arms.

"For me, it was an obvious choice as to who was going to swear me in," said Donnell. "I haven't seen my brother in four years, and for this opportunity to present itself, there was no other way I would have it done."

Lionel shared his brother's pride and enthusiasm and said he was honored for the opportunity to not only attend his brother's life-making decision but to actually conduct the ceremony itself.

"This is a great honor for me, for our Family, for him to let me do his re-enlistment," he said.

Both brothers will continue to serve in their respective units for the duration of the deployment, and both said they hope to visit home during their block leaves to share their experiences with the rest of their Family.

the deployment, and both said they hope to visit home during their block leaves to share their experiences with the rest of their Family.

CPT Lionel Macklin (left), a native of Ozark, AL, swears in his younger brother, SGT Donnell Macklin during a reenlistment ceremony at Camp Taji, northwest of Baghdad, Nov. 7. The brothers hadn't seen each other over the past four years due to service obligations but were reunited during Operation Iraqi Freedom 07-09. Lionel, the older brother, serves as a platoon leader with the 4th Infantry Division's A/3-4th Aviation Squadron, Combat Aviation Brigade, which is based out of Fort Hood, TX. Donnell serves as a forward signal support noncommissioned officer with B/52 Infantry Regiment "Hammer," which is based out of Hawaii. (Photo by SSG Guillermo Allen)

SSG Kristopher Verrett, a Fort Myers, FL native gives a "high five" to a boy from the Rashidyah Nahia during a patrol on Nov. 8. Verrett serves as a platoon sergeant with B/1-27 Infantry Regiment "Wolfhounds." (Photo by SGT Brad Willeford)

SSG Kristopher Verrett, a Fort Myers, FL native takes time to get to know, and share a few laughs with the children of the Rashidyah Nahia during a patrol on Nov. 8. Verrett serves as a platoon sergeant with B/1-27 Infantry Regiment "Wolfhounds." (Photo by SGT Brad Willeford)

SGT Ilalio Saufoi, of Pago Pago, Samoa, talks with children in the Istaqlal Qada on Nov. 9. Saufoi is an infantry team leader with A/1-27 Infantry Regiment "Wolfhounds" (Photo by SPC Stephen Miller)

SSG Matthew Weitz, a Flinstone, GA native, talks with two children during a patrol on Nov. 8 in the Rashidyah Nahia. Weitz serves as an infantry squad leader with B/1-27 Infantry Regiment "Wolfhounds." (Photo by SGT Brad Willeford)

MND-B engineers upgrade, expand site for fellow Soldiers

By 2LT Virginia Brickner
B/46th ECB (H), 926th Eng. Bde., MND-B

JOINT SECURITY STATION ISTAQLAL, Iraq – The relative comfort of main base camps in Iraq can make Soldiers forget that their fellow warriors are often stationed at remote combat outposts and joint security stations that have only the most basic of necessities.

In early August, Soldiers from 2nd Platoon, B/46 Engineer Combat Battalion (Heavy), based in Fort Rucker, AL, serving with Multi-National Division – Baghdad, set out from Camp Liberty and spent several months to make upgrades at JSS Istaqlal, north of Baghdad, to make the remote site a better place for Soldiers to live.

"The main focus of our mission was to provide much-needed quality of life upgrades for the Infantry Soldiers on the ground, as well as improving the operational capability of the unit as a whole," said SGT Peter Erikson, 46th ECB (H), squad leader and native of Des Moines, IA.

Soldiers constructed formwork in preparation of the placement of five concrete pads for future buildings, while a team of plumbers, led by SSG Ian Jones, a native of Newport, NY, built a gravity fed shower.

Elsewhere on the JSS, an electrical team led by SGT Michael Worman, who hails from Boise, ID, worked to safely ground electrical wires and cables on numerous generators and existing buildings.

Back at the main camp, Soldiers from B Co. busied themselves pre-fabricating all of the walls and trusses that would be used in the living quarters and the Combined Operations Center. Once all of the prefabricated materials and remaining Soldiers arrived at JSS Istaqlal, construction began in earnest.

"The living quarters were completed first and were the most direly needed. They were utilized less than 24 hours after the final nail was pounded," commented Jones.

Before the team was complete, fourteen additional buildings were built. A small group of engineers completed all ground excavation and concrete placement, while the rest of the platoon built all necessary walls and trusses at Camp Liberty.

The Combined Operations Center was the final building to be completed. This building allows both U.S. and Iraqi Security Forces to coordinate and plan security and stability missions as a team.

Due to the MND-B engineer's efforts, JSS Istaqlal has expanded to accommodate a much larger force.

SFC Vernon Alcorn, 2nd Plt. Platoon Sergeant, a native of Weiser, ID, was immensely proud of his highly-skilled and professional team of engineers.

"The Soldiers have surpassed my expectations," stated Alcorn. "Putting maximum effort into bettering the quality of life for their fellow Soldiers speaks volumes about their character. The message our Soldiers project is that we will get the job done, no matter what you throw at us. I am looking forward to our future missions, positive that our Soldiers will be successful."

Alcorn also states the superior carpentry skills and craftsmanship of the B Co. engineers was the reason the project was finished over a month ahead of schedule.

SPC Miles McFarland, carpenter, 46th Engineer Combat Battalion (Heavy), of Portland, OR; SPC Sean Cutis, of Toledo, OH; and SPC Brad McMahon, of Festus, MO, erect exterior walls for living quarter to house American Forces, Oct 7. (Photo by 2LT Virginia Brickner)

(Clockwise from top left) Carpenters PFC Chance Bailey, of Russellville, AR; and PFC Jason Steffes, who hails from Lincoln Park, MI, nail up fascia board as the sun sets in the distance.
(Photo by 2LT Virginia Brickner)

PVT Troy Emerson, 46th Engineer Combat Battalion (Heavy), of Bangor, Maine, uses a Bobcat to cut a trench for buried electrical cable while SSG Ryan Peltier, a squad leader, and a native of Baton Rouge, LA, ensures the job is done safely.
(Photo by 2LT Virginia Brickner)

SGT James Dill, 46th Engineer Combat Battalion (Heavy),of Dallas-Fort Worth, TX, drills wire holes in wall studs during pre-fabrication operations at Camp Liberty.
(Photo by 2LT Virginia Brickner)

Sasakwa, OK native, SPC Arthur Smith patrols through the Hussseniyah Nahia on Nov. 10, ensuring that residents in this northeast Baghdad town remain safe and free of terrorist activity. Smith is an infantryman with A/1-27 Infantry Regiment "Wolfhounds."
(Photo by SSG JB Jaso III)

San Diego native, SPC Eric Gonzalez communicates his unit's location on a tactical radio on Nov. 10, while military leaders conduct a meeting with Husseniyah Nahia Iraqi Police leaders. Gonzalez is a communications specialist with A/1-27 Infantry Regiment "Wolfhounds."
(Photo by SSG JB Jaso III)

Winchendon, MA native, SPC Garrett Simoneau maintains guard with an Iraqi Police officer on Nov. 10, ensuring that military and police leaders are safe during a security meeting improving the safety of residents in this Husseniyah Nahia town. Simoneau is an infantryman with A/1-27 Infantry Regiment "Wolfhounds." (Photo by SSG JB Jaso III)

SFC Dion Walker, of Louisville, KY, uses his magnified scope to maintain situational awareness while securing a Hussaniyah council meeting on Nov. 17. Walker is a platoon sergeant assigned to A/1-27 Infantry Regiment "Wolfhounds." (Photo by SPC John Ahn)

On Nov. 19 in the Istaqlal Qada, northeast of Baghdad, 1LT Joseph McCoy, a Lexington, NC native, discuses tactics with a 2nd Brigade, 1st Iraqi National Police Division officer in support of a weapons cache operation. McCoy is an infantry platoon leader with A/1-27 Infantry Regiment "Wolfhounds." (Photo by SPC Brian Pierce)

On Nov. 19, in the Istaqlal Qada northeast of Baghdad, SPC Jaymes Beebe maintains security while other soldiers in his squad search houses in support of weapon cache clearance operations. Beebe is an infantryman assigned to A/1-27 Infantry Regiment "Wolfhounds." (Photo by SPC Brian Pierce)

On Nov. 19, in the Istaqlal Qada northeast of Baghdad, 1LT Christopher McCoy, a Lexington, NC native, discusses tactics with 2nd Brigade, 1st Iraqi National Police Division officer in support of weapon cache clearance operation. McCoy is an infantry platoon leader for A/1-27 Infantry Regiment "Wolfhounds." (Photo by SPC Brian Pierce)

SGT Benjamin Dang, of West Orange, NJ, gives his undivided attention to a young child during a patrol to clear the Hussaniyah Nahia of weapons caches on Nov. 23. Dang is a combat medic assigned to A/1-27 Infantry Regiment "Wolfhounds." (Photo by SPC John Ahn)

Lancaster, CA native, SPC Jon Bothke; and Franklin, NC native, SGT Abraham Norton move towards a building in order to clear potential threats during a patrol in the Hussaniyah Nahia on Nov. 23. Bothke is a rifleman, and Norton is a team leader, both assigned to A/1-27 Infantry Regiment "Wolfhounds." (Photo by SPC John Ahn)

Lancaster, CA native, SPC Jon Bothke is followed by a large group of children during a dismounted patrol near a Hussaniyah Nahia primary school on Nov. 23. Bothke is a rifleman serving in A/1-27 Infantry Regiment "Wolfhounds." (Photo by SPC John Ahn)

(Left) PFC Jeff Arnold, of Cheney, WA, maintains security in the prone position as elements of his platoon conduct a search of the area in the Hussaniyah Nahia on Nov. 23. Arnold is an automatic rifleman assigned to A/1-27 Infantry Regiment "Wolfhounds." (Photo by SPC John Ahn)

(Right) Chesaning, MI native, SGT Gary Fox, maintains security in a grassy field in the Hussaniyah Nahia during a dismounted patrol on Nov. 23. Fox serves as a team leader in A/1-27 Infantry Regiment "Wolfhounds." (Photo by SPC John Ahn)

(Left) PFC Jeff Arnold, a Cheney, WA native, provides security for the Iraqi National Police as they conduct a quick meeting during a cache search mission in the Hussaniyah Nahia on Nov. 23. Arnold is an automatic rifleman assigned to A/1-27 Infantry Regiment "Wolfhounds." (Photo by SPC John Ahn)

On Nov. 24, SPC Richard Plasket, a Midlothian, TX native, shows off months of hard work during a strong man competition. Plasket is an infantryman for A/1-27 Infantry Regiment "Wolfhounds." (Photo by SPC Brian Pierce)

MAJ Patrick Aspland, of Fort Ann, NY, enjoys a cup of chai tea with a leading businessman in the Rashidiyah Nahia on Nov. 24. Aspland is the executive officer of 1-27 Infantry Regiment "Wolfhounds."
(Photo by SPC John Ahn)

On Nov. 29, SPC Jacob Atchinson, a South China, ME native, is greeted by one of Iraqs future leaders during mission in the Istaqlal Qada northeast of Baghdad. Atchinson is a forward observer for A/1-27 Infantry Regiment "Wolfhounds." (Photo by SPC Brian Pierce)

On Nov. 29, CPT Lucas Yoho, a Long Beach, CA native, enjoys a complementary snack during an important meeting with key leaders in the Istaqlal Qada northeast of Baghdad. Yoho commands A/1-27 Infantry Regiment "Wolfhounds." (Photo by SPC Brian Pierce)

(Left) The officers wait for the senior enlisted men of 1-27 Infantry Regiment "Wolfhounds" to break their huddle during the annual "Wolfhound Turkey Bowl," on Nov. 26. (Photo by SPC John Ahn)

(Right) Senior enlisted Soldiers of 1-27 Infantry Regiment "Wolfhounds" prepare a play during the annual "Wolfhound Turkey Bowl," on Nov. 26. (Photo by SPC John Ahn)

(Left) Portland, OR native, CPT William Lincoln, catches a one handed reception for a score during the annual "Wolfhound Turkey Bowl," on Nov. 26. Lincoln is the executive officer of B/52 Infantry Regiment "Hammer." (Photo by SPC John Ahn)

The officers of 1-27 Infantry Regiment "Wolfhounds" meet together and devise a game plan for the annual "Wolfhound Turkey Bowl," game that is played every year which pits the officers against the senior enlisted Soldiers. The game was played at a joint security station in the Istaqlal Qada. (Photo by SPC John Ahn)

Geneva, OH native, SSG Charles Bentley, rushes the officers' offensive line after the ball is snapped during the annual Thanksgiving Turkey Bowl, which matches senior enlisted against the officers of 1-27 Infantry Regiment "Wolfhounds." Bentley is a squad leader assigned to B/1-27 Infantry Regiment "Wolfhounds." (Photo by SPC John Ahn)

Seminole, FL native, 1LT Tommy Ryan runs through an open lane during a rush play on Nov. 26 during the annual "Wolfhound Turkey Bowl," which was played on a joint security station in the Istaqlal Qada. Ryan is a platoon leader assigned to A/1-27 Infantry Regiment "Wolfhounds." (Photo by SPC John Ahn)

The senior enlisted Soldiers of 1-27 Infantry Regiment "Wolfhounds" walk to the line of scrimmage after a huddle during the annual "Wolfhound Turkey Bowl," on Nov. 26. (Photo by SPC John Ahn)

1LT Joshua Kordela, of Owings Mills, MD, cracks in to a king crab leg during Thanksgiving Dinner on Nov. 27. Kordela is a platoon leader assigned to B/52 Infantry Regiment, 1-27 Infantry Regiment "Wolfhounds." (Photo by SPC John Ahn)

SSG Jonathan Castro, a Cape May, NJ native, tends to ham steaks in preparation for the Thanksgiving feast to be served to cavalry troopers. Castro is a cook for B/52 Infantry Regiment "Hammer," attached to 2-14 Cavalry Regiment. (Photo by MAJ Al Hing)

CPT Andrew Farina, of Franklin, NH; and CPT John Goldsworthy, of Richmond, WA, serve Thanksgiving dinner to Soldiers on a joint security station located in the Istaqlal Qada on Nov. 27. Farina is the commander of B/1-27 Infantry Regiment "Wolfhounds" and Goldsworthy leads the embedded local reconstruction team. (Photo by SPC John Ahn)

SGT James Snow, of Cape Coral, FL, gives a double *shaka* as he prepares a Thanksgiving meal for the Soldiers on a joint security station in the Istaqlal Qada on Nov. 27. Snow is a cook assigned to the 225th Brigade Support Battalion, 2nd Stryker Brigade Combat Team "Warrior." (Photo by SPC John Ahn)

El Paso, TX native, SPC Eric Esquivel enjoys Thanksgiving turkey at a joint security station in the Istaqlal Qada on Nov. 27. Esquivel is an infantryman assigned to B/1-27 Infantry Regiment "Wolfhounds." (Photo by SPC John Ahn)

Buffalo, NY native, SGT Edward Mara, takes pleasure in his Thanksgiving meal at a joint security station in the Istaqlal Qada on Nov. 27. Mara is a team leader assigned to B/1-27 Infantry Regiment "Wolfhounds." (Photo by SPC John Ahn)

Hot Springs, AR native, SPC Emmanuel Alvarez speaks with a young man in front of a primary school in the Rashidiyah Nahia on Nov. 28. Alvarez is an infantryman assigned to HHC/1-27 Infantry Regiment "Wolfhounds." (Photo by SPC John Ahn)

SSG Rizalino Corpuz, of Wahiawa, HI, delivers hams to the hungry Soldiers stationed on a joint security station in the Istiqlal Qada on Nov. 27. Corpuz is a cook assigned to the 225th Brigade Support Battalion, 2nd Stryker Brigade Combat Team "Warrior." (Photo by SPC John Ahn)

Henryetta, OK native, SPC Brian Barnett is approached by a young Iraqi cyclist during a dismounted patrol near the Rashidiyah marketplace in the Istaqlal Qada on Nov. 28. Barnett is a mortarman assigned to HHC/1-27 Infantry Regiment "Wolfhounds."
(Photo by SPC John Ahn)

Miami native, 1LT Kevin Strayhorn leads a patrol in the Rashidiyah Nahia while gathering atmospherics on Nov. 28. Strayhorn is the fire support officer for HHC/1-27 Infantry Regiment "Wolfhounds." (Photo by SPC John Ahn)

Bedford, IN native, SGT Eric Waddle pulls security during a dismounted patrol through a Rashidiyah Nahia neighborhood on Nov. 28. Waddle is a team leader assigned to HHC/1-27 Infantry Regiment "Wolfhounds." (Photo by SPC John Ahn)

El Paso, TX native, SSG Jesus Robles talks with students in front of a Rashidiyah Nahia middle school on Nov. 28. Robles is a section leader assigned to HHC/1-27 Infantry Regiment "Wolfhounds." (Photo by SPC John Ahn)

On Nov. 21, SSG Jesus Robles, an El Paso, TX native, takes a moment to smell the roses and appreciate the finer things in life during foot patrol in the Istaqlal Qada northeast of Baghdad. Robles is a squad leader for HHC/1-27 Infantry Regiment "Wolfhounds." (Photo by SPC Brian Pierce)

On Dec. 21, 2008, SSG Micheal Bird, of Bristol, CT, walks away from a Ministry of Defense hospital with a smile knowing that the Government of Iraq's Ministry of Health is going to make this hospital operational for the residents of the Istaqlal Qada. Bird is a combat medic with HHC/1-27 Infantry Regiment "Wolfhounds." (Photo by SPC Brian Pierce)

SSG Joseph Simplicano, a Kapolei, HI native, jokes around with two Iraqi boys Dec. 22, 2008 during Operation Gunslinger Riptide XX in the Istaqlal Qada. Simplicano serves as a squad leader with A/1-27 Infantry Regiment "Wolfhounds." (Photo by SGT Brad Willeford)

Wolfhounds, Iraqi National Police search for illegal weapons

By SSG JB Jaso III
1st Bn., 27th Inf. Regt., 2nd SBCT, 25th Inf. Div., attached to 3rd BCT, 4th Inf. Div., MND-B

ISTAQLAL, Iraq – Soldiers from A/1-27 Infantry Regiment "Wolfhounds" and members of the 2nd Brigade, 1st Iraqi National Police Division, weathered the high winds and cold winter temperatures in search of illegal weapons caches Dec. 22, 2008 during Operation Gunslinger Bonzai XXVI.

The search led the combined forces to the Hussaniyah town of the Istaqlal Qada, where they searched the open fields, empty compounds and historic cache locations for any signs of illegal weapons.

In the past, the Wolfhounds have found mortar rounds, improvised-explosive device making material and weapons in the area that could've been used to harm Iraqi Security Forces, civilians and Coalition Forces.

Even though the Wolfhounds didn't find any weapons this day, the combined operation was a giant step toward continued security in Iraq, said 1LT Tommy Ryan, a native of Seminole, FL, who serves as a platoon leader.

"They were very knowledgeable and offered lots of advice," said Ryan, in reference to the NPs who assisted in the operation. "We were able to gather intelligence that was crucial to understanding a high-value individual in our area of operations."

The mission served to strengthen the bonds of the combined team for future operations. "All patrols are now combined," said CPT Timothy Hall, the battalion's assistant operations officer, who is a native of Fairfax Station, VA, in talking about the frequency of patrols the Wolfhounds are conducting with the Iraqi Security Forces. "We are ensuring that the conditions are set for the ISF to assume full responsibility of the area's security."

However, noted Ryan, working together is nothing new for the MND-B Soldiers. "We have been conducting combined operations with the Iraqi Security Forces since day one," Ryan stated. "The practice hasn't changed."

Members of the 2nd Brigade, 1st Iraqi National Police Division, prepare to enter a building Dec. 22, 2008 that is known to be a historic illegal weapons cache location during Operation Gunslinger Bonzai XXVI in the Istaqlal Qada. Soldiers from A/1-27 Infantry Regiment "Wolfhounds" have been assisting the NPs in securing this northeast Baghdad region. (Photo by SSG JB Jaso III)

1LT Tommy Ryan, a Seminole, FL native, discusses plans with an Iraqi National Police officer Dec. 22, 2008 in the Istaqlal Qada during Operation Gunslinger Bonzai XXVI. Ryan is an infantry platoon leader with A/1-27 Infantry Regiment "Wolfhounds." (Photo by SSG JB Jaso III)

SSG Ahki Savea, of Pago Pago, Samoa, searches for illegal weapons during Operation Gunslinger Bonzai XXVI on Dec. 22, 2008, in the Husseniyah town of the Istaqlal Qada. Savea is an infantry squad leader with A/1-27 Infantry Regiment "Wolfhounds."
(Photo by SSG JB Jaso III)

SGT Ahki Saufoi, a Pago Pago, Samoa, native, receives orders from 1LT Tommy Ryan, a Seminole, FL native, at the beginning of Operation Gunslinger Bonzai XXVI, a mission focused at identifying and removing illegal weapons caches in the Hussaniyah town of the Istaqlal Qada on Dec. 22, 2008. Saufoi is an infantry team leader, and Ryan is an infantry platoon leader. Both are assigned to A/1-27 Infantry Regiment "Wolfhounds." (Photo by SSG JB Jaso III)

PFC Ryan Thompson a Coben, IL native, maintains security during Operation Gunslinger Bonzai XXVI, a mission focused at identifying and removing illegal weapon caches in the Hussaniyah town of the Istaqlal Qada Dec. 22, 2008. Thompson is an infantryman with A/1-27 Infantry Regiment "Wolfhounds." (Photo by SSG JB Jaso III)

SSG Jose Perez, of Bayamon, Puerto Rico, takes a moment to give a thumbs up after destroying several illegal ordnance caches found Dec. 26, 2008 during Operation Havoc Bonzai VIII in the Istaqlal Qada. Perez serves as a combat medic with HHC/1-27 Infantry Regiment "Wolfhounds." (Photo by SGT Brad Willeford)

SGT Mark Davenport, a Wimberly, TX native, pulls security Dec. 22, 2008 prior to Operation Gunslinger Riptide XX, an operation focused at capturing a high value individual in the Istaqlal Qada, northeast of Baghdad. Davenport serves as a medic with A/1-27 Infantry Regiment "Wolfhounds." (Photo by SGT Brad Willeford)

CPT Lucas Yoho, a Long Beach, CA native, puts his foosball skills to the test Dec. 22, 2008 during Operation Gunslinger Riptide XX in the Istaqlal Qada. Yoho commands A/1-27 Infantry Regiment "Wolfhounds." (Photo by SGT Brad Willeford)

SFC Dion Walker, a Louisville, KY native; and PFC Corey Show, an Irwin, PA native, take aim to clear an open field Dec. 22, 2008 during Operation Gunslinger Riptide XX in the Istaqlal Qada, northeast of Baghdad. Walker serves as a platoon sergeant and Show as an infantryman with A/1-27 Infantry Regiment "Wolfhounds." (Photo by SGT Brad Willeford)

1LT John Busuego, an Ontario, CA native; and 1LT Erik Wiesehan, a Canby, OR native, serenade guests with traditional Christmas songs in the Rashidiyah Nahia, located northeast of Baghdad, during a Christmas barbeque hosted by a local sheik Dec. 26, 2008. Busuego and Wiesehan are members on the embedded Local Reconstruction Team of 1-27 Infantry Regiment "Wolfhounds." (Photo by SPC John Ahn)

SFC Dion Walker, a Louisville, KY native, teaches an Iraqi boy how to pound fists Dec. 22, 2008 during Operation Gunslinger Riptide XX in the Istaqlal Qada, northeast of Baghdad. Walker serves as a platoon sergeant with A/1-27 Infantry Regiment "Wolfhounds."
(Photo by SGT Brad Willeford)

Midlothian, TX, native, PFC Richard Plasket sometimes uses his boot as a breach kit while conducting "Riptide" missions in the Istaqlal Qada. Plasket is an infantryman with A/1-27 Infantry Regiment "Wolfhounds." (Photo by SSG JB Jaso III)

PFC Carlos Fisher, of Laramie, WY, smiles at a child inside a Husseniyah town marketplace on Dec. 24, 2008 while ensuring the safety of residents in this northeast Baghdad community. Fisher is an infantryman with A/1-27 Infantry Regiment "Wolfhounds." (Photo by SSG JB Jaso III)

PFC Kenneth Bledsoe, a native of Kansas City, MO, patrols through a marketplace in the Istaqlal Qada's Husseniyah town on Dec. 24, 2008 ensuring the safety of children and shoppers. Bledsoe is an infantryman with A/1-27 Infantry Regiment "Wolfhounds." (Photo by SSG JB Jaso III)

SSG Curtis Johnson, of Waterport, NY, directs his squad of infantrymen through a Husseniyah marketplace on Dec. 24, 2008 ensuring the safety of residents conducting business in this northeast Baghdad area. Johnson is an infantry squad leader with A/1-27 Infantry Regiment "Wolfhounds." (Photo by SSG JB Jaso III)

LTC Richard Wilson, of Brooklyn, NY, speaks with a Ministry of Education representative prior to the opening of the Imam Al Sadiq School in Istaqlal Qada's Husseniyah town on Dec. 24, 2008. Wilson commands the 1-27 Infantry Regiment "Wolfhounds." (Photo by SSG JB Jaso III)

Upland, CA native, CPL Steven Randall, maintains security in front of a barber shop in Istaqlal Qada's Husseniyah town on Dec. 24, 2008 allowing residents in this northeast Baghdad community to conduct their business without fear of criminal or terrorist activity. Randall is a psychological operations specialist with the 1230th Tactical Psychological Operations Detachment, 315th Tactical Psychological Operations Company. (Photo by SSG JB Jaso III)

Providence, RI native, SPC Dustin Lapolla patrols along a busy business district in the Husseniyah town of the Istaqlal Qada on Dec. 24, 2008 allowing residents to conduct business without fear of criminal or terrorist activity. Lapolla is a forward observer with A/1-27 Infantry Regiment "Wolfhounds." (Photo by SSG JB Jaso III)

Wolfhounds capture criminal, then celebrate second Christmas in Iraq

by SSG JB Jaso III
1-27 Infantry Regiment, 2nd SBCT, 25th Inf. Div., attached to 3rd BCT, 4th Inf. Div., MND-B

As a fifteen-month deployment comes to a close for Wolfhound Soldiers they remain focused at continuing their mission, even during the holiday season.

The Wolfhounds, who are more formally known as 1-27 Infantry Regiment "Wolfhounds," celebrated their second consecutive Christmas in Iraq, but the holiday celebrations didn't keep Soldiers from HHC from capturing one of the area's most wanted terrorists.

An early Christmas morning combined operation with the 2nd Brigade, 1st Iraqi National Police Division resulted in the capture of an individual associated with improvised explosive devices and planning attacks with suicide vest bombs stated the Wolfhounds assistant intelligence officer, 1LT Joseph McKinney, of Stafford, VA.

"Taking this type of individual off the streets is really helping to increase the security for the Iraqi people as they continue to take the reins on security in their cities," continued McKinney. "This country is better off without these criminals in the population and it is a huge step for the Iraq Security Forces to have them help run the operations, with Coalition Forces, as these criminals are pulled off the streets."

The capture resulted in a personal Christmas morning wake-up call for the Wolfhound platoon leader responsible for the capture. "It was surprising in a good way; a pleasant surprise," said CPT Pete Walther, of Mililani, HI, who was awakened in person by the 3rd BCT, 4th Inf. Div. commander, COL John Hort, who flew to Joint Security Station Istaqlal on Christmas morning to congratulate Walther and his Soldiers for the capture. "It was really great for him to come and congratulate my men and I."

Hort personally congratulated the sleepy Soldiers who only had a couple hours of sleep after their Christmas wish of capturing this criminal was granted. He awarded commander coins for excellence to the Soldiers directly involved with the capture and to other outstanding Soldiers with A/1-27 Infantry Regiment.

"I was excited to get my first coin and to meet COL Hort," said SPC Benjamin Raulerson who was selected by his PSG to be recognized by the brigade commander. Raulerson is an A Co. infantryman and Jacksonville, AL native.

Once all the congratulations were completed, the Wolfhounds began their second consecutive Christmas feast of turkey, ham, crab legs and shrimp, which was cooked and prepared by the four cooks assigned to the battalion. "The food was great," stated Shreveport, LA native and A Co. infantryman, PFC Timothy Moreland. "The cooks did a great job!"

Not only did the Wolfhounds enjoy their Christmas dinner, but their Iraqi counterparts, enjoyed the feast as well. "It was vitally important to include our Iraqi Security Force partners in our holiday season celebrations and feast to reinforce unity of effort in all we do as a team; just as we currently do in combined operations today," stated the Wolfhound Commander and Brooklyn, NY native, LTC Richard "Flip" Wilson. "This is another historical period in the country of Iraq where our bonding together represents the everlasting hope of mankind and the forward progress of security by, with and through each Iraqi Force in this area."

As Christmas came to a close, the tired Wolfhounds reflected on their accomplishments and celebrations of the day as Wilson stated that "we celebrated the very fact that we were together today securing and protecting the population of this area."

SPC Kristopher Curtis, of Effingham, IL, prepares a Christmas feast for Soldiers from the 1-27 Infantry Regiment "Wolfhounds" at a joint security station in the Istaqlal Qada on Dec. 25, 2008. Curtis is a cook with the 225th Brigade Support Battalion. (Photo by SSG JB Jaso III)

COL John Hort speaks with Soldiers from A/1-27 Infantry Regiment "Wolfhounds" before awarding them a commander's coin for excellence at a joint security station in the Istaqlal Qada Dec. 25, 2008. Hort commands the 3rd BCT, 4th Infantry Division.
(Photo by SSG JB Jaso III)

CPT Omololu Makinde, of Columbus, OH, serves Christmas dinner to SPC Andrew Liu, of Fremont, CA, at a joint security station in the Istaqlal Qada on Dec. 25, 2008. Makinde is the Iraqi Security Forces coordinator and Liu is a military intelligence analyst. Both are assigned to the 1-27 Infantry Regiment "Wolfhounds." (Photo by SSG JB Jaso III)

SPC Joel Wright, of Waterloo, IA, serves himself shrimp to top-off his Christmas dinner at a joint security station in the Istiqlal Qada on Dec. 25, 2008. Wright is a reconnaissance scout with HHC/1-27 Infantry Regiment "Wolfhounds." (Photo by SSG JB Jaso III)

SPC Steven Williams, of Oklahoma City, OK, enjoys his Christmas dinner with members of the 2nd Brigade, 1st Iraqi National Police Division that celebrated the holiday with U.S. Soldiers at a joint security station in the Istaqlal Qada on Dec. 25, 2008. Williams is an infantryman with A/1-27 Infantry Regiment "Wolfhounds." (Photo by SPC Brian Pierce)

PFC Kenneth Bledsoe, a native of Kansas City, MO, maintains security in a busy Husseniyah town marketplace on Dec. 24, 2008. Bledsoe is an infantryman with A/1-27 Infantry Regiment "Wolfhounds." (Photo by SSG JB Jaso III)

SSG Ahki Savea, of Pago Pago, Samoa, dons some holiday cheer while he provided additional security at Baghdad's International Zone on Dec. 24, 2008. Savea is an infantry squad leader with A/1-27 Infantry Regiment "Wolfhounds." (Photo by SPC Stephen Miller)

On Dec. 24, 2008, SPC Clinton Coey, a Morning Sun, IA native, secures the outer perimeter of a house while the other members of his team search inside for illegal weapon caches during Operation Havoc Bonzai VI in the Istaqlal Qada, northeast of Baghdad. Coey is a reconnaissance scout with HHC/1-27th Infantry Regiment "Wolfhounds." (Photo by SPC Brian Pierce)

SGT Joseph Miramontez, a Balch Springs, TX native, takes a long drink of water during a Christmas Day football game between Soldiers from different units assigned to Forward Operating Base War Eagle located in Baghdad. Miramontez is an infantry team leader assigned to B/1-27 Infantry Regiment "Wolfhounds." (Photo by SPC John Ahn)

SPC Sebastian Klui, a Lincoln Park, NJ native, takes time to call his wife in Hawaii on Christmas Day from FOB War Eagle in Baghdad. Klui is an infantry team leader assigned to B/1-27 Infantry Regiment "Wolfhounds." (Photo by SPC John Ahn)

SPC Dustin Smith, a Dyersburg, TN native, works out with weights at the Forward Operating Base War Eagle gym located in Baghdad on Christmas Day. Smith is an infantryman assigned to B/1-27 Infantry Regiment "Wolfhounds." (Photo by SPC John Ahn)

Soldiers of B/1-27 Infantry Regiment "Wolfhounds" enjoy a Christmas Day meal at the FOB War Eagle dining facility in Baghdad. (Photo by SPC John Ahn)

Christmas in Iraq

CPT Peter Walther, a Mililani, HI native; and CPT Brian Sweigart, a San Antonio native, observe UH-60 helicopters land to pick them up Dec. 26, 2008 for Operation Havoc Bonzai VIII in the Istaqlal Qada. Walther serves as a reconnaissance platoon leader and Sweigart is the commander for HHC/1-27 Infantry Regiment "Wolfhounds." (Photo by SGT Brad Willeford)

SGT Joshua Thorner, a Troutdale OR, native, briefs several members of the Iraqi emergency response unit Dec. 26, 2008 before Operation Havoc Bonzai VIII, an operation focused at identifying and removing illegal caches in the Istaqlal Qada. Thorner is a reconnaissance scout for the HHC/1-27 Infantry Regiment "Wolfhounds." (Photo by SGT Brad Willeford)

SPC Joshua Fulgham (right), of Inverness, MS; SPC Joel Wright, of Waterloo, IA; and SPC Ronald Jarrard, of Holly Hill, FL, wait to board UH-60 utility helicopters before Operation Havoc Bonzai VIII in the Istaqlal Qada Dec. 26., 2008. Fulgham, Wright, and Jarrard are reconnaissance scouts with HHC/1-27 Infantry Regiment "Wolfhounds."
(Photo by SGT Brad Willeford)

SPC Joel Wright, a Waterloo, IA native, gazes out the window of a UH-60 utility helicopter Dec. 26, 2008 during Operation Havoc Bonzai VIII in the Istaqlal Qada. Wright is a reconnaissance scout with the HHC/1-27 Infantry Regiment "Wolfhounds." (Photo by SGT Brad Willeford)

(Left) 1LT Corey Harris, a Phoenix native, sets up a Christmas tree at the house of a local leader in the Rashidiyah Nahia, located northeast of Baghdad, during a Christmas barbeque Dec. 26, 2008. Harris is the civil affairs officer from Detachment 2, B/403 Civil Affairs Brigade, assigned to 1-27 Infantry Regiment "Wolfhounds." (Photo by SPC John Ahn)

(Right) MAJ Patrick Aspland, a Fort Ann, NY native, is visited by a little Santa Claus during a Christmas barbeque at a Rashidiyah Nahia council member's home, located northeast of Baghdad, Dec. 26, 2008. Aspland is the executive officer for 1-27 Infantry Regiment "Wolfhounds." (Photo by SPC John Ahn)

(Below) SSG Kellen Hansen, a Los Angeles native, takes time to enjoy a cup of Chai tea during a Christmas celebration at a local Iraqi leader's home in the Rashidiyah Nahia, located northeast of Baghdad, Dec. 26, 2008. Hansen is a reconnaissance scout assigned to HHC/1-27 Infantry Regiment "Wolfhounds." (Photo by SPC John Ahn)

In the Rashidiyah Nahia, Soldiers from B/1-27 Infantry Regiment "Wolfhounds" check out a building to see what it would take to refurbish it for future use Dec. 28, 2008.
(Photo by SGT Brad Willeford)

SSG David Ni, a San Gabriel, CA native, watches as his men establish security during a patrol in the Rashidiyah Nahia of northern Baghdad Dec. 28, 2008. Ni is an infantry squad leader with B/52 Infantry Regiment attached to 1-27 Infantry Regiment "Wolfhounds."
(Photo by SGT Brad Willeford)

1LT Louis Cascino, a Windsor, NY native, walks with members of the Rashidiyah Nahia's Emergency Response Unit during a combined patrol of the neighborhood Dec. 28, 2008. Cascino serves as an infantry platoon leader for B/1-27 Infantry Regiment "Wolfhounds."
(Photo by SGT Brad Willeford)

Wolfhounds Train Iraqi Police

By SFC JB Jaso III
1st Bn., 27th Inf. Regt., 2nd SBCT, 25th Inf. Div.

ISTAQLAL, Iraq – As U.S. Forces in Iraq transition all of their operations by, with, and through the Iraqi Security Forces, "Wolfhound" Soldiers continue their efforts in ensuring the Iraqi Police in the Istaqal Qada are ready to protect the residents in this northeast Baghdad region.

The Soldiers from A/1-27 Infantry Regiment "Wolfhounds" conducted tactical training with their Iraqi Police counterparts Dec. 29, 2008.

"We taught them entering and clearing rooms and buildings, sensitive site exploitation and searching individuals," said SPC Joshua Lewis, of Dayton, OH, who was one of the instructors for training. "We taught them the tips and tricks you need to know.

"It was important to teach them to ensure that they stay alive and safe." Lewis continued.

When planning the course, the NCOs responsible for the training had to overcome the language barrier when trying to implement the finer details of this training.

"It was difficult trying to explain the way to do the maneuvers in simple words," said SGT Tony Macias, an infantry squad leader and native of Wauchula, FL, who used an interpreter to train the Iraqi Police. "It went good though. We are trying to teach them the basics."

An Iraqi Police officer agreed that the training went well: "It was good training," said the IP officer. "I look forward to more training by the Coalition Forces."

As the Iraqi Security Forces take an even stronger responsibility for the security of their nation, the "Wolfhounds" remain dedicated to ensuring that their counterparts are prepared.

"The security environment in Istaqlal Qada and northern Adhamiyah fosters coordination between members of the Al Zahour IP Security District, the 6th Emergency Response Unit Battalion and the 2nd Brigade, 1st Iraqi National Police Division," said LTC Richard "Flip" Wilson, of Brooklyn, NY, the "Wolfhound" commander. "We, as a security force, on any given day honorably serve our respective nations with great discipline and teamwork."

SPC Jeremy Mingle, of Homer, NY, instructs an Iraqi Police officer how to safely and properly enter a building during a training session Dec. 29, 2008 in the Istaqlal Qada. Mingle is an infantryman with A/1-27 Infantry Regiment "Wolfhounds." (Photo by SFC JB Jaso III)

SGT Tony Macias, of Wauchula, FL, demonstrates to an Iraqi Police officer the proper way to maintain control of a suspected criminal during a training session held in the Istaqlal Qada Dec. 29, 2008. Macias is an infantry team leader with A/1-27 Infantry Regiment "Wolfhounds." (Photo by SFC JB Jaso III)

SGT Eberardo Solis, of Mission, TX, explains the importance of properly searching an individual for weapons and illegal items to Iraqi Police officers during a training session Dec. 29, 2008 in the Istaqlal Qada. Solis is an infantry team leader with A/1-27 Infantry Regiment "Wolfhounds." (Photo by SFC JB Jaso III)

SGT Tony Macias, of Wauchula, FL, provides guidance and feedback to an Iraqi Police officer upon the completion of a training session held in the Istaqlal Qada Dec. 29, 2008. Macias is an infantry team leader with A/1-27 Infantry Regiment "Wolfhounds." (Photo by SFC JB Jaso III)

SPC David Mensink, of Carl Junction, MO, evaluates Iraqi Police officers conduct a thorough search of an individual during a training session held in the Istaqlal Qada Dec. 29, 2008. Mensink is an infantryman with A/1-27 Infantry Regiment "Wolfhounds." (Photo by SFC JB Jaso III)

SSG David Ni, a San Gabriel, CA native, hands out candy to children during a patrol in the Rashidiyah Nahia Dec. 28, 2008. Ni is an infantry squad leader with B/52 Infantry Regiment attached to 1-27 Infantry Regiment "Wolfhounds." (Photo by SGT Brad Willeford)

PFC Dylan Guell, an Eden, WI native, talks with several children during a patrol in the Rashidiyah Nahia Dec. 28, 2008. Guell serves as an infantryman with B/1-27 Infantry Regiment "Wolfhounds."
(Photo by SGT Brad Willeford)

SPC Kyle Lauth, of Calverton, NY, sips an energy drink in preparation for a dismounted patrol through the Hussaniyah town of the Istaqlal Qada Dec. 29, 2008. Lauth is an infantryman with A/1-27 Infantry Regiment "Wolfhounds."
(Photo by ***SFC*** JB Jaso III - Congratulations on the promotion!)

PFC Christopher Mola, of Valparaiso, IN, patrols through the streets of Istaqlal Qada's Hussaniyah town Dec. 29, 2008 ensuring the safety of the residents in this northeast Baghdad community. Mola is an infantryman with A/1-27 Infantry Regiment "Wolfhounds." (Photo by SFC JB Jaso III)

1LT Tommy Ryan, of Seminole, FL, gives a thumbs-up to children in the Hussaniyah town of the Istaqlal Qada Dec. 22, 2008 after explaining to them the importance of reporting the location of illegal weapons during Operation Gunslinger Bonzai XXVI. Ryan is an infantry platoon leader with A/1-27 Infantry Regiment "Wolfhounds."
(Photo by SSG JB Jaso III)

Belleville, IL native, SPC William Coyner dismounts a Stryker armored vehicle and inspects the surrounding area for enemy activity Dec. 19, 2009 in the Rashidiyah Nahia. Coyner is a mortarman assigned to HHC/1-27 Infantry Regiment "Wolfhounds."
(Courtesy Photo)

Long Beach, CA native, CTR1(SW) Nathan Francis problem solves Electronic Warfare issues plaguing Stryker armored vehicles at Joint Security Station Istaqlal, Dec. 29, 2008. Francis is the electronics warfare officer assigned to 1-27 Infantry Regiment "Wolfhounds."
(Courtesy Photo)

SSG Philip Snook, a native of Eldersburg, MD, is surrounded by students from al Yasmin School, northwest of Baghdad, Jan. 3. IP officers and Soldiers assigned to B/52 Infantry Regiment "Hammer," handed out pens, pencils, crayons and paper to children at the school. Snook is an infantry team leader assigned to Hammer Co.
(Photo by SPC Charles Joseph)

SSG Brian Wallace, of Warrenton, MO, patrols through a Husseniyah marketplace Jan. 3, 2009, ensuring the continued safety of residents in this northeast Baghdad community. Wallace is an infantry squad leader with A/1-27 Infantry Regiment "Wolfhounds." (Photo by SFC JB Jaso III)

SGT Adam Brown, of Annandale, VA, hands out school supplies with Iraqi Police at the al Yasmin School, northwest of Baghdad, Jan. 3. The IP officers and Soldiers assigned to B/52 Infantry Regiment "Hammer," handed out pens, pencils, crayons and paper to the children. Brown is an infantryman assigned to Hammer Co.
(Photo by SPC Charles Joseph)

SPC Christopher Cumbie, of Broken Arrow, OK, provides security during a market assessment through the Rashidiyah Nahia of the Istaqlal Qada Jan. 3, 2009. Cumbie is an infantryman with A/1-27 Infantry Regiment "Wolfhounds." (Photo by SFC JB Jaso III)

SGT Timothy Miller, of Circleville, OH, conducts navigation training with officers from the Al Zahour Iraqi Police Station, in Istaqlal Qada's Husseniyah town Jan. 4, 2009 to help prepare them to assume full responsibility of security in this northeast Baghdad region. Miller is an infantry team leader with A/1-27 "Wolfhounds." (Photos by SFC JB Jaso III)

1LT Aaron Kletzing, of Chicago, meets another Chicago Bears fan while patrolling through a marketplace in the Rashidiyah Nahia, Jan. 3, 2009. Kletzing is the fire support officer for A/1-27 Infantry Regiment "Wolfhounds." (Photo by SFC JB Jaso III)

Helena, MT native, 1SG Todd Ballenger, takes a seat on the ramp of a Stryker armored vehicle on Jan. 4, 2009 at Joint Security Station Sadr City, during a combat patrol. Ballenger is the first sergeant of B/1-27 Infantry Regiment "Wolfhounds." (Courtesy Photo)

At the Al Zahour Iraqi Police Station in Istaqlal Qada's Husseniyah town, SSG Christopher Bolwell, of Boca Raton, FL, teaches an Iraqi boy the fundamentals of throwing a baseball, in hopes of spreading the baseball fever to Iraq, Jan. 4, 2009. Bolwell is an infantry squad leader with A/1-27 Infantry Regiment "Wolfhounds." (Photos by SFC JB Jaso III)

SSG Christopher Bolwell, of Boca Raton, FL, teaches an Iraqi boy the importance of the pitching windup Jan. 4, 2009. at the Al Zahour Iraqi Police Station in Istaqlal Qada's Husseniyah town. Bolwell is an infantry squad leader with A/1-27 Infantry Regiment "Wolfhounds." (Photos by SFC JB Jaso III)

Chesapeake, VA native, SGT William Heuisler plays a video game during some down time at FOB War Eagle, located in Baghdad, Jan 4, 2009. Heuisler is a stryker armored vehicle commander with B/1-27 Infantry Regiment "Wolfhounds." (Courtesy Photo)

Zion, IL native, PFC Scott Paugel secures an Iraqi Army checkpoint behind a concrete barrier during a halt while conducting dismounted patrols in Baghdad Jan. 6, 2009. Paugel is an automatic rifleman assigned to B/1-27 "Wolfhounds." (Courtesy Photo)

SPC Nick Pata, of New City, NY, is joined with several children, while he patrols through the Abu Ghuraib Qada, Jan. 6, 2009. Pata is an infantryman with C/1-27 Infantry Regiment "Wolfhounds." (Courtesy Photo)

Evans, CO native SGT Trevor Sullivan; and Dallas native, SPC Jason Ellis walk through an urban Baghdad neighborhood, with Iraqi Army Soldiers, Jan. 6, 2009 during a combined patrol. Sullivan and Ellis are infantrymen with B/1-27. (Courtesy Photo)

Millbrook, AL native, SSG Brian Smith inspects Iraqi Army checkpoints in Baghdad in order to secure the population from attacks, Jan. 6, 2009. Smith is an infantry squad leader assigned to B/1-27 "Wolfhounds." (Courtesy Photo)

Fresno, CA native, SPC Joseph Lorenz takes a moment to pose with Iraqi children during a combat patrol in Baghdad Jan. 7, 2009. Lorenz is a combat medic assigned to B/1-27 Infantry Regiment "Wolfhounds." (Courtesy Photo)

Boston native, SPC Steven Murphy patrols, with Iraqi Army Soldiers, through Baghdad during a combined dismounted patrol Jan. 7, 2009. Murphy is a radio telephone operator assigned to B/1-27 Infantry Regiment "Wolfhounds." (Courtesy Photo)

SGT Cameron Carney, of Midland, TX; and SPC Sebastian Klui, of Jersey City, NJ, pose with their mustaches near a Stryker armored vehicle in Baghdad Jan. 7, 2009. Carney and Klui are infantrymen assigned to B/1-27 "Wolfhounds." (Courtesy Photo)

SPC Nick Pata, of New City, NY, enjoys a cup of chai while patrolling through the Abu Ghuraib Qada, Jan. 6, 2009. Pata is an infantryman with C/1-27 Infantry Regiment "Wolfhounds." (Courtesy Photo)

Brawley, CA native, CPT Richard Pace gives an Iraqi Army officer directions on handbill dissemination aimed to capture high value individuals throughout Baghdad Jan. 7, 2009. Pace is an infantry platoon leader assigned to B/1-27 Infantry Regiment "Wolfhounds." (Courtesy Photo)

SGT Jason Stalnaker, a native of Newark, OH, pulls security behind a concrete barrier in Sadr City as he and his Iraqi Army counterparts patrolled the area. Stalnaker servers with B/1-27 Infantry Regiment "Wolfhounds." (Photo by SGT Whitney Houston)

SPC Franklin Walz, a native of Kansas City, MO, patrols the streets of Sadr City Jan. 8, 2009 with an Iraqi Army counterpart from the 45th Brigade, 11th Division. Due to the new security agreement, Coalition Forces must operate directly with the IA operatives to do any kind of operation. Walz serves with B/1-27 Infantry Regiment "Wolfhounds." (Photo by SGT Whitney Houston)

LAMBRIGHT

An Iraqi Army soldier from the 24th Brigade, 6th IA Division, and SPC Travis Clabaugh, an Edwards Air Force Base, CA native, pull security during a short halt on a combined patrol in the Abu Ghraib District, west of Baghdad, Jan. 9, 2009. Increased combined patrols have made this neighborhood safer for the families to live. Clabaugh is an infantryman assigned to C/1-27 Infantry Regiment "Wolfhounds". (Photo by 1LT Christopher Smith)

An Iraqi Army soldier with the 45th Brigade, 11th Iraqi Army Division, pulls security Jan. 8, 2009 while on patrol in Sadr City with Coalition Forces counterpart, SGT Sean Cheker, a Chicago native, serving with B/1-27 Infantry Regiment "Wolfhounds." (Photo by SGT Whitney Houston)

SPC David Dempsey, of Travelers Rest, SC, talks with a member of the 2nd Brigade, 1st Iraqi National Police Division Jan. 13, 2009, during a combined operation, focused at identifying and removing illegal weapons caches in the Istaqlal Qada Jan. 13. Dempsey is a mortarman with HHC/1-27 Infantry Regiment "Wolfhounds." (Photo by SGT Brad Willeford)

Soldiers from HHC/1-27 Infantry Regiment "Wolfhounds," board a UH-60 Blackhawk helicopter following the completion of a combined operation, with the 2nd Brigade, 1st Iraqi National Police Division Jan. 13, 2009. The operation was focused at identifying and removing illegal weapon caches in the Istaqlal Qada.
(Photo by SGT Brad Willeford)

SSG Jonathan Castro, of Cape May, NJ, prepares dinner for soldiers at JSS Hor al Bosh, Jan. 13, 2009. Castro is a cook assigned to 225th Brigade Support Battalion, attached to B/52 Infantry Regiment "Hammer." (Photo by SFC JB Jaso III)

SGT Christopher Clay, of Marysville, CA, reviews the company's property book, accounting for organizational and stay-behind equipment in the finally weeks remaining in the deployment, Jan. 14, 2009. Clay is a supply clerk with HHC/1-27 Infantry Regiment "Wolfhounds." (Photo by SFC JB Jaso III)

MAJ Patrick Aspland (left), of Fort Ann, NY; and MAJ Michael Zernickow, of Concord, CA, enjoy the new addition to the battalion's tactical operations center, a cardboard cut-out of SSG Donald Fabian, "Abu Shouarb," of Miami, FL, Jan. 14, 2009 at JSS Istaqlal. SSG Fabian was the "Wolfhounds" TPT Team Leader for 11-months and became known as the "father of the mustache." Aspland is the executive officer and Zernickow is the operations officer for the 1-27 Infantry Regiment "Wolfhounds."
(Photo by SFC JB Jaso III)

SPC Jose Nunez, of Chicago, digs up an illegal weapons cache, Jan. 17, 2009 in the Hor al Bosh area of the Taji Qada. Nunez is an infantryman with A/1-27 Infantry Regiment "Wolfhounds."
(Photo by SFC JB Jaso III)

SGT Gary Fox, of Chesaning, MI, briefs the soldiers of HHC/1-27 Infantry Regiment "Wolfhounds" of prohibited items prior to inspecting their bags, Jan. 14, 2009 at Camp Taji. Fox is an infantry team leader with A/1-27 IN. (Photo by SFC JB Jaso III)

SPC Joshua Fulgham reviews his deployment checklist, ensuring that he completes all the required briefs needed to return to the U.S., Jan. 14, 2009 at Camp Taji. Fulgham is a reconnaissance scout with HHC/1-27 Infantry Regiment "Wolfhounds." (Photo by SFC JB Jaso III)

SSG Jose Perez, of Bayamon, Puerto Rico, repacks his duffle bags Jan. 14, 2009 at Camp Taji, following a U.S. customs inspection, required to pack bags inside containers headed back to Schofield Barracks. Perez is a combat medic with HHC/1-27 Infantry Regiment "Wolfhounds." (Photo by SFC JB Jaso III)

SPC Luke Butler, of Gooding, ID, unpacks his duffle bags that he plans to pack in unit containers to be inspected by customs agents, Jan. 14, 2009 at Camp Taji. Butler is a reconnaissance scout with HHC/1-27 Infantry Regiment "Wolfhounds." (Photo by SFC JB Jaso III)

SGT Gary Fox, of Chesaning, MI, checks bags that will be packed into containers, ensuring that illegal items are not transported back to Hawaii, Jan. 14, 2009 at Camp Taji. Fox is an infantry team leader with A/1-27 Infantry Regiment "Wolfhounds." (Photo by SFC JB Jaso III)

SPC Blake Kerr, of Bonnots Mill, MO, dumps the contents of his duffle bags to be inspected by military working dogs and customs agents, Jan. 14, 2009 at Camp Taji, ensuring that illegal items are not transported to the U.S. Kerr is a supply clerk for HHC/1-27 Infantry Regiment "Wolfhounds." (Photo by SFC JB Jaso III)

SGT Jared Bruyneel, of San Bernardino, CA, inspects the bags of SPC Tyrell McCurbin, of Brooklyn, NY, during redployment operations, ensuring that illegal items are not transported back to Hawaii, following a 15-month deployment in Iraq, Jan. 14, 2009 at Camp Taji. Bruyneel is an infantry team leader with A/1-27 Infantry Regiment "Wolfhounds." McCurbin is a intelligence analyst with the "Wolfhounds."
(Photo by SFC JB Jaso III)

SPC Donald Lane, of Kansas City, MO, hands SPC Ryan Thompson, of Cobden, IL, new nightstands that will be installed into containerized housing units (CHUs) that will house the 1-112 Infantry Regiment "Strong" (PANG), Jan. 14, 2009 at Camp Taji. Lane and Thompson are infantrymen with A/1-27 Infantry Regiment "Wolfhounds."
(Photo by SFC JB Jaso III)

SFC Benjamin Chapman, of Barboursville, WV, talks with several Iraqi boys as they came running up to ask his name Jan. 16, 2009 during a combined patrol with the 2nd Brigade, 1st Iraqi National Police Division, in the Istaqlal Qada. Chapman is a reconnaissance platoon sergeant with HHC/1-27 Infantry Regiment "Wolfhounds." (Photo by SGT Brad Willeford)

PFC David Newkirk, of Newburgh, NY, makes a new friend during a combined patrol with the 2nd Brigade, 1st Iraqi National Police Division, in the Istaqlal Qada Jan. 16, 2009. The patrol was assisting the Government of Iraq in continuing to secure this northeast Baghdad region. Newkirk serves as a reconnaissance scout with HHC/1-27 Infantry Regiment "Wolfhounds." (Photo by SGT Brad Willeford)

SPC Jacob Bruno, of Randolph, MN, prepares to put a bandage on an Iraqi boy's hand after he scraped it Jan. 16, 2009. Bruno was on a combined patrol with policemen from 2nd Brigade, 1st Iraqi National Police Division, in the Istaqlal Qada. Bruno is a combat medic with HHC/1-27 Infantry Regiment "Wolfhounds." (Photo by SGT Brad Willeford)

PFC James Lohuis a Glassboro, NJ native, patrols alongside an Iraqi Army soldier from the 3rd Battalion, 45th Brigade, 11th IA Division, during a combined dismounted patrol focused at learning patrolling techniques in an urban area of Baghdad Jan. 16, 2009. Lohuis is an infantryman assigned to B/1-27 Infantry Regiment "Wolfhounds." (Photo by SPC John Ahn)

PVT Christopher McGinnis an Eldon, MO native, walks along concrete barriers during a dismounted patrol in Baghdad Jan. 16, 2009. McGinnis is an automatic rifleman assigned to B/1-27 Infantry Regiment "Wolfhounds." (Photo by SPC John Ahn)

SPC Jose Nunez, of Chicago, detects for additional illegal weapons caches, Jan. 17, 2009, after locating one in the Hor al Bosh area of the Taji Qada. Nunez is an infantryman with A/1-27 Infantry Regiment "Wolfhounds." (Photo by SFC JB Jaso III)

SSG Alaster Anderson, of Carrollton, MS, enjoys his time working hand-in-hand with the 2nd Brigade, 1st Iraqi National Police Division during a combined patrol through Baghdad's Subak Sur community, Jan. 18, 2009. Anderson is an infantry squad leader with B/52 Infantry Regiment, attached to A/1-27 Infantry Regiment "Wolfhounds." (Photo by SPC John Ward)

SSG George Canfill, of La Mesa, CA, is given a Bavaria beer while conducting a combined patrol in Baghdad's Subak Sur community Jan. 18, 2009. Canfill returned the beer to the generous gentlemen. Canfill is the human intelligence team leader for A/1-27 Infantry Regiment "Wolfhounds." (Photo by SFC JB Jaso III)

SGT Joseph McCoy, of Branson, MO, sits and talks with a young boy, while patrolling through the Thalba community of Baghdad's Adhamiyah security district, Jan. 21, 2009. McCoy is an infantry team leader with B/52 Infantry Regiment. (Photo by SPC John Ward)

CSM Karl Morgan (right) presents newly promoted SGM Timothy Foster, of Lynderborough, NH, with his new rank during a ceremony at Joint Security Station Istaqlal Jan. 20, 2009. Foster serves as the operations sergeant major for 1-27 Infantry Regiment "Wolfhounds." (Photo by SGT Brad Willeford)

MAJ Michael Zernickow, of Concord, CA, congratulates SGM Timothy Foster, of Lynderborough, NH, at Foster's promotion ceremony at Joint Security Station Istaqlal Jan. 20, 2009. (Photo by SGT Brad Willeford)

SPC Keagan Mooring, of Portland, TX, patrols through the Thalba area of Baghdad's Adhamiyah security district, Jan. 21, 2009. Mooring is an infantryman with B/52nd Infantry Regiment. (Photo by SPC John Ward)

SSG Felipe Torres, of Oxnard, CA; and 1LT Corey Harris, of Phoenix begin to reconstruct the Ministry of Defense hospital in the Istaqlal Qada, Jun. 22. Torres is a mortarman and Harris is the civil affairs officer with 1-27 "Wolfhounds." (Photo by SFC JB Jaso III)

SSG Brian Wallace, of Warrenton, MO, maintains security outside of the Imam Al Sadik School in the Husseniyah town Jan. 22, 2009, during an election polling site visit. Wallace is an infantry squad leader with A/1-27 "Wolfhounds." (Photo by SFC JB Jaso III)

SGT Michael Rios, of Sebastian,TX, follows a member of the 2nd Brigade, 1st Iraqi National Police Division into a Hussaniyah Nahia school, Jan. 24, 2009 that is being used as a polling center during Iraq's provincial elections. (Photo by SFC JB Jaso III)

SPC Michael Ortiz, of Riverhead, NY, patrols through a Hussaniyah Nahia school Jan. 24, 2009, while conducting an inspection of polling sites with the Al Salam Iraqi Police. Ortiz is an infantry team leader with A/1-27 "Wolfhounds." (Photo by SPC John Ward)

SGT Christopher Vaccaro (left), of St. Francisville, LA; and SPC Corey Show, of Irwin, PA maintain security in the Hussaniyah Nahia, along with Al Salam Iraqi Police, Jan. 24, 2009. Vaccaro and Show are mortarmen with A/1-27 "Wolfhounds." (Photo by SPC John Ward)

SGM Timothy Foster, of Lynderborough, NH, maintains security while military leaders assess a compact water treatment unit in Istaqlal Qada's Husseniyah town. Foster is the operations SGM for 1-27 Infantry Regiment "Wolfhounds." (Photo by SFC JB Jaso III)

SGT Christopher Vaccaro, of St. Francisville, LA, meets children in the Hussaniyah Nahia, Jan. 24, 2009. Vaccaro is a mortarman with A/1-27 Infantry Regiment "Wolfhounds." (Photo by SPC John Ward)

SGT Joseph Baker, of Blanco, TX, writes the grid coordinates of a Hussaniyah Nahia school that is being used as a provincial election polling center, Jan. 24, 2009. Baker is a forward observer with A/1-27 Infantry Regiment "Wolfhounds." (Photo by SFC JB Jaso III)

SPC Jehad Mahmoud, of Sagamore Hills, OH patrols through the Hussaniyah Nahia, while members of the Al Salam Iraqi Police, disseminate handbills, Jan. 25, 2009. Mahmoud is infantryman with A/1-27 Infantry Regiment "Wolfhounds." (Photo by SPC John Ward)

SPC John Ward, of El Paso, TX, holds a baby while patrolling through the Boob al Sham community Jan. 26, 2009. Ward is a combat medic with B/52 Infantry Regiment, attached to A/1-27 Infantry Regiment "Wolfhounds." (Courtesy Photo)

SPC Charles Vanderhalf, of Poulsboro, WA, maintains rear security while conducting a dismounted patrol through Baghdad's Adhamiyah Security District, along with members of the Boob al Sham Iraqi Police, Jan. 26, 2009. Vanderhalf is a mortarman with A/1-27 Infantry Regiment "Wolfhounds." (Photo by SPC John Ward)

SGT Joseph McCoy, of Branson, MO, maintains security, while the Boob al Sham Iraqi Police disseminate flyers, notifying residents of this northern Adhamiyah security district of how to contact them in case of an emergency, Jan. 26, 2009. McCoy is an infantry team leader with B/52 Infantry Regiment, attached to A/1-27 Infantry Regiment "Wolfhounds." (Photo by SPC John Ward)

SPC Michael Weber, of Greely, CO, maintains security, providing security to families in Baghdad's Adhamiyah security district, Jan. 26, 2009. Weber is an infantryman with B/52 Infantry Regiment, attached to A/1-27 Infantry Regiment "Wolfhounds."
(Photo by SPC John Ward)

SGT Joseph McCoy, of Branson, MO, patrols through the Boob al Sham community, Jan. 26, 2009, ensuring the continued security of residents in this northern Baghdad community. McCoy is an infantry team leader with B/52 Infantry Regiment, attached to A/1-27 Infantry Regiment "Wolfhounds." (Photo by SPC John Ward)

SPC Travis Craig, of Chicora, PA, maintains security while leaders from the 2nd Brigade, 1st Iraqi National Police Division inspect an election polling center Jan. 27, 2009, in the Istaqlal Qada. Craig is an infantryman with B/52nd Infantry Regiment, attached to B/1-27 Infantry Regiment "Wolfhounds." (Photo by SFC JB Jaso III)

Ramsey, NJ native, SPC Sebastian Klui; and Midland, TX native, SGT Cameron Carney take positions over the Tigris River in Baghdad Jan. 28, 2009, during operations clearing water treatment facilities of insurgent threats during an escalated stance of security due to impending provincial elections in Iraq. Klui and Carney are infantry team leaders with B/1-27 Infantry Regiment "Wolfhounds." (Courtesy Photo)

Fresno, CA native, SSG Ernesto Rodriguez; and Melbourne, FL native, SSG John White question and positively identify a local national working at a water treatment facility located in Baghdad Jan. 28, 2009, during operations to secure the area from criminal threats ensuing from Iraq's provincial elections. Rodriguez is an infantry squad leader with B/1-27 Infantry Regiment "Wolfhounds." (Courtesy Photo)

MILLER

صيدليه

SGT Joseph Baker (left), of Blanco, TX; and SPC Christopher Cumbie, of Midvale, UT, provide security during BG Robin Swan's visit to Husseniyah town Jan. 22, 2009. Baker and Cumbie are infantrymen with A/1-27 Infantry Regiment "Wolfhounds." Swan is the deputy commanding general for the 4th Infantry Division and MND-B. (Photo by SFC JB Jaso III)

CPL Eugene Dimauro, of Ambler, PA, prepares to board a Stryker armored vehicle, Jan. 27, 2009, in the Hussaniyah Nahia. Dimauro is an infantry team leader with A/1-27 Infantry Regiment "Wolfhounds." (Photo by SGT Brad Willeford)

SSG Daniel Ramirez, of San Diego, watches his Soldiers during a combined patrol with the Al Salam Iraqi Police, Jan. 27, 2009 in the Hussaniyah Nahia. Ramirez is an infantry squad leader with A/1-27 Infantry Regiment "Wolfhounds." (Photo by SGT Brad Willeford)

SPC Jaymes Beebe, of Kansas City, KS, maintains security outside a polling site in Hussaniyah on Jan. 27, 2009, while members of the Al Salam Iraqi police ensure the location is safe for the upcoming provincial elections. Beebe serves as an infantryman with A/1-27 Infantry Regiment "Wolfhounds." (Photo by SFC JB Jaso III)

SGT Joseph Baker, of Blanco, TX, maintains security along with members of the 2nd Brigade, 1st Iraqi National Police Division, immediately following a roadside bomb that caused no injuries in the Hussaniyah Nahia, Jan. 29, 2009. Baker is a forward observer with A/1-27 Infantry Regiment "Wolfhounds." (Photo by SFC JB Jaso III)

1LT Tommy Ryan, of Seminole, FL, explains the details of a roadside bomb that exploded in the Hussaniyah Nahia, Jan. 29, 2009, to LTC Richard "Flip" Wilson, of Brooklyn, NY. Ryan is an infantry platoon leader with A Co., while Wilson commands the 1-27 Infantry Regiment "Wolfhounds." (Photo by SFC JB Jaso III)

1LT Tommy Ryan, of Seminole, FL, talks to Soldiers in his platoon, praising them for their dedication and hard work, following a roadside bomb in the Hussaniyah Nahia, Jan. 29, 2009. The Soldiers in Ryan's platoon, along with members of the 2nd Brigade, 1st Iraq National Police Division, secured the area, ensuring no injuries were sustained by the explosion. (Photo by SFC JB Jaso III)

SSG Rodriquez Cruell (left), of Greenville, NC; and SPC Christopher Mola, of Valparaiso, IN, maintain security following a roadside bomb that exploded in the Hussaniyah Nahia, Jan. 29, 2009. Cruell and Mola are infantrymen with A/1-27 Infantry Regiment "Wolfhounds." (Photo by SFC JB Jaso III)

SGT Scott Geiszler, of Albany, OR; and a member of the 2nd Brigade, 1st Iraqi National Police Division maintain security at a polling center in the Istaqlal Qada, Jan. 30, 2009. Geiszler is an infantryman with B/52nd Infantry Regiment. (Photo by SFC JB Jaso III)

SPC Travis Craig, of Chicora, PA, patrols through the Fahama neighborhood of Baghdad's Adhamiyah security district, Jan. 30, 2009, less than 24-hours prior to Iraq's provincial elections. Craig is an infantryman with B/52nd Infantry Regiment. (Photo by SFC JB Jaso III)

(Above) SPC Jaymes Beebe, of Kansas City, KS, maintains security, Jan. 30, 2009, during a combined patrol with the 2nd Brigade, 1st Iraqi National Police Division prior to the upcoming provincial elections in the Hussaniyah Nahia. Beebe serves as an infantryman with A/1-27 Infantry Regiment "Wolfhounds." (Photo by SGT Brad Willeford)

A member of the 2nd Brigade, 1st Iraqi National Police Division leaps across a canal during a combined patrol Jan. 31, 2009, in the Rashidiyah Nahia, with soldiers from HHC/1-27 Infantry Regiment "Wolfhounds." (Photo by Brad Willeford)

SSG Gregory Antoine, of Brooklyn, NY, patrols through a historic weapons cache location, during Operation Havoc Shorebreak VI, Jan. 27, 2009, in Gumayra. The combined operation, with the 2nd Brigade, 1st Iraqi National Police Division, protected the upcoming provincial elections from enemy or terrorist activity. Antoine is an infantry squad leader with B/52 Infantry Regiment "Hammer." (Photo by SFC JB Jaso III)

SGT Douglas Urias, of Houston, takes time to talk with a member from the 2nd Brigade, 1st Iraqi National Police Division Jan. 30, 2009, prior to a combined patrol in the Hussaniyah Nahia. Urias is an infantry team leader with A/1-27 Infantry Regiment "Wolfhounds." (Photo by SGT Brad Willeford)

SPC Travis Craig, of Chicora, PA, maintains security in Baghdad's Adhamiyah security district, allowing the 2nd Brigade, 1st Iraqi National Police Division to establish a polling site for Iraq's provincial elections. Craig is an infantryman with B/52 Infantry Regiment, attached to A/1-27 Infantry Regiment "Wolfhounds." (Photo by SFC JB Jaso III)

Zion, IL native, PFC Scott Paugel searches for explosive making materials for use against election polling stations located throughout Baghdad Jan. 31, 2009. Paugel is an infantryman with B/1-27 Infantry Regiment "Wolfhounds." (Courtesy Photo)

Amarillo, TX native, PFC Billy Hill takes cover behind a large date tree and provides security as members of his squad search residential homes Jan. 31, 2009 in Baghdad. Hill is an infantryman with B/1-27 Infantry Regiment "Wolfhounds." (Courtesy Photo)

Evans, CO native, SGT Trevor Sullivan takes a brief meeting with his squad leader to receive instructions for a follow-on mission to search a different area of Baghdad and inspect a possible improvised explosive device Jan. 31, 2009. Sullivan is an infantry team leader with B/1-27 Infantry Regiment "Wolfhounds." (Courtesy Photo)

Dallas native, SPC Jason Ellis maintains security in an affluent Baghdad neighborhood during Iraq's provincial elections Jan. 31, 2009. Ellis is an infantryman with B/1-27 Infantry Regiment "Wolfhounds." (Courtesy Photo)

Clarkston, FL native, 1LT Kenneth Jura is given a high five by a young Iraqi boy during patrols searching residential areas for materials that could be used to disrupt Iraq's provincial elections, Jan 31, 2009. Jura is an infantry platoon leader with B/1-27 Infantry Regiment "Wolfhounds." (Courtesy Photo)

Albany, OR native, PFC Dillon Jayne; and Belton, MO native, SPC Franklin Walz walk through a Baghdad neighborhood during security operations focused at securing Iraq's provincial elections Jan. 31, 2009. Jayne and Walz are infantrymen with B/1-27 "Wolfhounds." (Courtesy Photo)

Amarillo, TX native, PFC Billy Hill; and Dallas native, SPC Jason Ellis search an expansive garden in Baghdad during operations focused at ensuring a safe provincial election Jan. 31, 2009. Hill and Ellis are infantryman with B/1-27 "Wolfhounds." (Courtesy Photo)

Yuma, AZ native, SGT Joseph Galvan searches residential areas throughout Baghdad in order to maintain security during elections on Jan 31, 2009. Galvan is an infantry team leader with B/1-27 Infantry Regiment "Wolfhounds." (Courtesy Photo)

Residents from the Husseniyah town of the Istaqlal Qada show their ink-stained fingers, proving that they participated in Iraq's Provincial Elections, Jan. 31, 2009. Soldiers from the 1-27 Infantry Regiment "Wolfhounds" assisted the Iraqi Security Forces in providing additional security, allowing Iraqis to vote without fear of criminal or terrorist activity. (Photos by SFC JB Jaso III)

Soldier In Focus: 1SG Eugene Dydasco

Today's Soldier in Focus is 1SG Eugene Dydasco, an Agana Heights, Guam, native, currently working as a senior enlisted leader at Joint Security Station Istaqlal in northeastern Baghdad.

Dydasco serves with A/1-27 Infantry Regiment "Wolfhounds," 2nd Stryker Brigade Combat Team "Warrior," 25th Infantry Division, currently attached to 3rd Brigade Combat Team, 4th Infantry Division, Multi-National Division – Baghdad, out of Schofield Barracks, Hawaii.

He said that watching Soldiers grow into future leaders is one of the things that makes him proud to serve.

"Serving my country is a way of giving back to the community," said the 17-year Army veteran.

(Photo by SPC Brian Pierce)

Wolfhounds Run to the Finish Line

By SFC JB Jaso III
1-27 Infantry Regiment "Wolfhounds"

ISTAQLAL, Iraq – As a 15-month deployment comes to a close, several "Wolfhounds" are running against time to make it to the finish. Soldiers assigned to the 1-27 Infantry Regiment "Wolfhounds" have formed a club that motivates the Soldiers to run laps around Joint Security Station Istaqlal, northeast of Baghdad, in an effort to fight to the finish.

"This was a spur of the moment idea" stated CPT Brian Sweigart, of San Antonio, on how the club formed. "It was an idea that was thought of when we were sitting around the table. We set an achievable goal in our short time remaining in Iraq," continued Sweigart, who was one of the original five runners that started counting laps on Nov. 11.

Sweigart, who commands HHC, said that 127 laps around JSS Istaqlal was a good number, as it represents the unit.

"We have a large area to run," stated SGT Timothy Miller, of Circleville, OH and infantry team leader in A Co. "Each lap is .75 of a mile long."

If finishing more than 95-miles in the last three months in the deployment wasn't difficult enough for some, the terrain was another added challenge.

"More than 50 percent of the course is covered with gravel and large rocks, which makes it difficult to run on," continued Miller, who was also one of the original club members. "It's like running in the sand on the beach."

The challenge of running 127 laps around the JSS was completed first by SFC Nathanial Hanrahan, of Broad Brook, CN, who is an infantryman with A Co.

Not only has Hanrahan completed the 127 laps, but he also recently completed 170 laps, resulting in over 127-miles ran since Nov. 11.

"The 127 lap club is a positive thing out here," stated Miller. "It has motivated a lot of people to get out there and run; it helped me get out there more than I would have."

CPT Omololu Makinde, of Columbus, OH, completes another lap around the rocky terrain at Joint Security Station Istaqlal Jan.19, getting closer to reaching his goal of 127 laps. Makinde is the Iraqi security forces coordinator for 1-27 Infantry Regiment "Wolfhounds." (Photo by SFC JB Jaso III)

The original five members of the 127 lap club forms at Joint Security Station Istaqlal to begin their run to the finish on Nov. 11. The runners are assigned to 1-27 Infantry Regiment "Wolfhounds." (Photo by SPC John Ahn)

WIESEHAN

SPC Jesse Barnes, of Sidney, NY, stands guard at the entrance of a home suspected of hiding illegal weapons, Feb. 1, 2009 in the Rashidiyah Nahia, during a combined operation with members of the 2nd Brigade, 1st Iraqi National Police Division. Barnes is an infantryman with B/52 Infantry Regiment "Hammer." (Photo by SGT Brad Willeford)

SPC Bryan Landers, of Muscle Shoals, AL, pulls security at the front of a building, during Operation Havoc Bonzai XV, a combined operation aimed at finding illegal weapons Feb. 1, 2009, in the Rashidiyah Nahia, with members of the 2nd Brigade, 1st Iraqi National Police Division. Landers is an infantryman with B/1-27 Infantry Regiment "Wolfhounds." (Photo by SGT Brad Willeford)

SPC Douglas Byers, of Grinnell, IA, digs in search of weapons caches, with the help of a military working dog, during Operation Havoc Bonzai XV, a mission aimed at finding illegal weapons Feb. 1, 2009 in the Rashidiyah Nahia. Byers serves as a infantryman with B/1-27 Infantry Regiment "Wolfhounds." (Photo by SGT Brad Willeford)

(Left) SFC James Burciaga (right), of West Covina, CA; and SSG Rodriguez Cruell, of Greenville, SC, review their map of the Hussaniyah area Feb. 3, 2009, during a combined air assault mission with members of the 2nd Brigade, 1st Iraqi National Police Division. Burciaga, an infantry platoon sergeant, and Cruell, an infantry squad leader, are both with A/1-27 Infantry Regiment "Wolfhounds." (Photo by SGT Brad Willeford)

(Right) SGT Jeremiah Sullivan, of St. Joseph, MI, leaps off of a downed fence after using it to cross a canal during a combined air assault operation with members of the 2nd Brigade, 1st Iraqi National Police Division during Operation Gunslinger Bonzai XXX, Feb. 3, 2009 in the Hussaniyah Nahia. Sullivan is an infantry team leader assigned to A/1-27 Infantry Regiment "Wolfhounds." (Photo by SGT Brad Willeford)

(Left) Soldiers with A/1-27 Infantry Regiment board a UH-60 Blackhawk helicopter after completing Operation Gunslinger Bonzai XXX Feb. 3, 2009 in the Hussaniyah Nahia. The combined air assault operation was conducted with members of the 2nd Brigade, 1st Iraqi National Police Division.
(Photo by SGT Brad Willeford)

1SG Eugene Dydasco, of Agana Heights, Guam, prepares the Soldiers of A/1-27 Infantry Regiment "Wolfhounds," for an awards ceremony at JSS Istaqlal, Feb. 3, 2009.
(Photo by SFC JB Jaso III)

SPC Nicholas Snyder displays an Iraqi sign of respect to CPT Lucas Yoho, of Long Beach, CA, after being award the Army Commendation Medal during an awards ceremony on Feb. 3, 2009. Snyder is a human intelligence analyst and Yoho commands A/1-27 Infantry Regiment "Wolfhounds." (Photo by SFC JB Jaso III)

SGT Vern Smith is awarded an Army Achievement Medal during an awards ceremony held at JSS Istaqlal on Feb. 3, 2009. Smith is a mortarman with A/1-27 Infantry Regiment "Wolfhounds." (Photo by SFC JB Jaso III)

During an awards ceremony at JSS Istaqlal, Feb. 3, 2009, CPT Lucas Yoho, of Long Beach, CA, praises the soldiers of A/1-27 Infantry Regiment "Wolfhounds," for their dedication and teamwork during their 15-month deployment in Iraq. Yoho commands A Co.
(Photo by SFC JB Jaso III)

SGT George Davis leads his men during a combined cache clearance operation with members of the 2nd Brigade, 1st Iraqi National Police Division Feb. 7, 2009 in the Rashidiyah Nahia. Davis, of Buena Vista, CO, serves as an infantry team leader with B/52 Infantry Regiment "Hammer." (Photo by SGT Brad Willeford)

1LT Louis Cascino, an infantry platoon leader, B/1-27 Infantry Regiment "Wolfhounds," calms a donkey after he and his team rescued it from a drainage ditch. Cascino, from Windsor, NY, partnered with members of the 2nd Brigade, 1st Iraqi National Police Division, Feb. 7, 2009 during a combined cache clearance operation in the Rashidiyah Nahia. (Photo by SGT Brad Willeford)

SGT George Davis (left), of Buena Vista, CO; SSG Davis Ni (center), of San Gabriel, CA; and 1LT Louis Cascino, of Windsor, NY, decide where to start searching during a combined cache clearance operation with officers from the 2nd Brigade, 1st Iraqi National Police Division Feb. 7, 2009 in the Rashidiyah Nahia, northeast of Baghdad. All are infantrymen with B/1-27 Infantry Regiment "Wolfhounds." (Photo by SGT Brad Willeford)

PFC Matthew McMahon, an infantryman with A/1-27 Infantry Regiment "Wolfhounds," uses a metal detector to search for illegal weapons caches in northeast Baghdad Feb. 9, 2009. McMahon, a native of Littleton, CO, partnered with Al Salaam Iraqi Police officers during the combined operation focused on ridding the Hussaniyah Nahia of enemy weapons.
(Photo by SFC JB Jaso III)

Cheney, WA native, PFC Jeffrey Arnold, an infantryman, assigned to A/1-27 Infantry Regiment "Wolfhounds," maintains security at the entrance of a house suspected of storing illegal weapons Feb. 9, 2009, while on a joint Iraqi Security Force operation in northeast Baghdad.
(Photo by SFC JB Jaso III)

SPC Corey Johnson, of Pleasant Prairie, WI, maintains security on top of the Al Salaam Iraqi Police Station, overlooking a mosque in the Hussaniyah town of the Istaqlal Qada, Feb. 10, 2009. Johnson is an infantryman with A/1-27th Infantry Regiment "Wolfhounds."
(Photo by SFC JB Jaso III)

SSG Andrew Lambright maintains security outside of a National Police patrol base in Istaqlal Qada's Hussaniyah town on Feb. 10, 2009. Lambright is an infantry team leader with A/1-27 Infantry Regiment "Wolfhounds." (Photo by SFC JB Jaso III)

Denson overcomes odds, challenges to serve in Iraq

by SFC JB Jaso III
1st Bn., 27th Inf. Regt., 2nd SBCT, 25th Inf. Div.

Overcoming challenges and defying the odds is nothing new to SPC David Denson.

The 37-year-old infantryman with A/1-27 Infantry Regiment "Wolfhounds," has spent his life doing what others said that he couldn't do.

Those people said this Oceanside, CA native, couldn't surf the Southern Californian waves with the professionals. However, Denson dedicated his young life to surfing. He surfed hard and began his professional surfing career riding for the Hobie, Life's a Beach, Body Glove, and Rip Curl surf teams.

"I busted by butt," Denson said, in regards to the effort it took to surf at the professional level. "I wasn't Tom Curren, but I got to where I wanted to be."

Where he wanted to be was in surfing's Mecca: Hawaii.

Denson moved to the island of Kauai in 1991, where he surfed the crystal clear Hawaiian waters and adapted to the culture there.

Just months after moving to Kauai, Denson learned that the perfect tropical Hawaiian weather could quickly change. In September 1992, Hurricane Iniki destroyed his home and took all of his belongings.

The hurricane's destruction was so significant that Denson felt the need to do everything he could to assist his new community. "It doesn't take much to be a part of a community," Denson said. "My parents taught me to be a good guy and do what is right."

The right thing for him was to temporarily give up surfing to help rebuild homes and volunteer at the local Salvation Army.

During this time, Denson learned the unique and treasured Hawaiian cultures and traditions. He said he learned how native Hawaiians have overcome challenges and the true meaning of aloha.

Assisting him in learning these new traditions and cultures was Hawaiian-native and great-great-great-granddaughter of Kalaimamahu, half-brother to Kamehameha I, Leiliwin Kalei Mahuiki.

Mahuiki and Denson fell in love and married in September 1995.

While they raised their four girls, Denson researched and became an activist for Hawaiian rights. He defended land that was owned by native Hawaiians on Kauai's north shore in 1999, which a prominent businessman attempted to purchase.

"I assisted in researching the issues that the Hawaiians believed in," stated Denson. "I wanted to make my Family proud and be a good role model for my kids."

He made his family proud, not only protecting native Hawaiian rights, but also participating in youth sports and anti-drug programs.

In 2001, Denson would be faced with another challenge.

SPC David Denson, of Hanalei, HI, is an infantryman with A/1-27 Infantry Regiment "Wolfhounds."
(Photo by SFC JB Jaso III)

While driving his pregnant wife on Kauai, Denson was involved in a vehicle accident that caused his wife to have a cerebral hemorrhage. Later, his wife passed away – along with their unborn child.

"I was devastated," Denson said after the accident. "I learned how fast life can change and what's important.

"Nobody thought I would be able to raise my kids," the now single father said. Eight years later, those people are impressed how hard he has worked to raise his four daughters, ages 7 through 12.

The girls have become fluent in their native Hawaiian language. The oldest has received a full scholarship to the Kamehameha Schools; and, of course, they all surf.

"They're a lot like their mom," Denson said about his daughters. "They have a lot of aloha for people.

"I didn't think I would come back from Leiliwin's death," Denson said.

In 2006, Denson knew he had to continue his selfless service to those in need.

"Spreading freedom to another country is similar to sticking up for the Hawaiians," Denson said upon joining the Army.

Denson proved those individuals that doubted his ability to complete basic combat training wrong when he graduated and deployed to Iraq.

Denson is currently serving as an infantryman in Iraq's Istaqlal Qada with the "Wolfhounds." For the last 14-months, he assisted the Iraqi Security Forces in securing the northern Baghdad region. This additional security has allowed the Government of Iraq to provide much-needed essential services, such a medical aid, education, water and electricity to the residents.

"We're trying to make a difference in people's lives," stated Denson. "We're sharing a little bit of Aloha while we connect with the population, letting them know we're here for them."

"It does well knowing the local culture, just like when I moved to Hawaii," Denson said upon arriving in Iraq. "It goes a long way learning a little Arabic."

Not only has Denson connected well with the local populace, he has also connected very well with the younger Soldiers in his squad. "The guys call me 'dad,'" Denson said. "I enjoy working with the young Soldiers and watching them grow."

"He is a person everyone goes to as a mentor," said, SGT Nathan Taylor, Denson's squad leader, who is a Greenville, MS native. "He is respected, and everyone knows he has a wealth of knowledge because of his pervious experiences."

Denson enjoys seeing the younger Soldiers, whom are nearly 20 years younger than him, stepping up from wherever they are from and seeing them achieve their goals and aspirations.

As Denson completes his 15-month tour in Iraq, he looks back at the accomplishments that he and his unit have accomplished and looks forward to spending time with his girls.

"I tell my girls that terrorists don't tend to the needs of the people," Denson said. "We did the right thing here, making the time to do the right missions. I'm proud to be a Wolfhound!"

SPC David Denson, of Hanalei, HI, patrols the Husseniyah town of the Istaqlal Qada, Sept. 14. Denson was assisting the Iraqi Security Forces in maintaining the security in this northeast Baghdad community. Denson is a infantryman with A/1-27 Infantry Regiment "Wolfhounds." (Photo by SFC JB Jaso III)

U.S. ARMY

Soldiers from 1-27 Infantry Regiment "Wolfhounds" and "1-112 Infantry Regiment "Strong" stand side-by-side during the Transfer of Authority ceremony at Camp Taji, Feb. 22, 2009. The "Wolfhounds" transferred authority of the Istaqlal Qada and northern Adhamiyah Security District to "Strong." (Photo by SFC JB Jaso III)

LTC Richard "Flip" Wilson, of Brooklyn, NY; and CSM William Hain, of Reno, NV, case the battalion's colors, officially ending the 1-27 Infantry Regiment "Wolfhounds" 15-month tour in Iraq on Feb. 22, 2009 at Camp Taji. (Photo by SFC JB Jaso III)

LTC Richard "Flip" Wilson, of Brooklyn, NY; and CSM William Hain, of Reno, NV, case the battalion's colors during a Transfer of Authority ceremony held at Camp Taji, Feb. 22, 2009, officially relinquishing their area of operations to the 1-112 Infantry Regiment "Strong." (Photo by SFC JB Jaso III)

LTC Richard "Flip" Wilson, of Brooklyn, NY, reflects on the 1-27 Infantry Regiment "Wolfhounds" 15-month deployment in northern Baghdad during a Transfer of Authority ceremony held at Camp Taji, Feb. 22, 2009. (Photo by SFC JB Jaso III)

SSG Mario Amaro, of Miami, passes the time by watching a movie before flying out of Camp Taji to Baghdad International Airport on Feb. 17, 2009. Amaro serves as an infantry squad leader with C/1-27 Infantry Regiment "Wolfhounds" (Photo by SPC Brian Pierce)

SPC Brandon Baricuatro (left), of Frisco, TX; and SPC Robert Hatala, of Ottawa, IL, await their flight from Baghdad International Airport to Kuwait City on Feb. 18, 2009. Baricuatro and Hatala are reconnaissance scouts with HHC/1-27 Infantry Regiment "Wolfhounds." (Courtesy Photo)

Soldiers from 1-27 Infantry Regiment "Wolfhounds" fly from Baghdad International Airport to Kuwait City inside an Air Force C-17 Globemaster on Feb. 18, 2009. (Courtesy Photo)

SPC Robert Rasely, of Bangor, PA, passes SPC Wilfred Ferris, of Fairfield, ME, a rucksack as they unload bags from a C-17 Globemaster at Kuwait City on Feb. 18, 2009. Rasely is a communications specialist and Ferris is a forward observer. Both are assigned to HHC/1-27 Infantry Regiment "Wolfhounds." (Photo by SPC Brian Pierce)

SPC Ryan Maxfield, of Howell, MI, reads a book to pass the time while waiting a flight from Kuwait City to the United States, Feb. 19, 2009. Maxfield is a combat medic with HHC/1-27 Infantry Regiment "Wolfhounds." (Photo by SPC Brian Pierce)

1SG Timothy Jackson, of Groton, NY, obtains accountability of personnel on the chartered civilian flight from Kuwait City to Hickam Air Force Base, in Honolulu, HI, on Feb. 20, 2009. Jackson is the first sergeant for HHC/1-27 Infantry Regiment "Wolfhounds." (Courtesy Photo)

SPC Corey Johnson, of Pleasant Prairie, WI, plays video games while awaiting a flight from Baghdad International Airport to Kuwait City on Feb. 24, 2009. Johnson spent nearly three days in Baghdad awaiting transportation to Kuwait. Johnson is an infantryman with A/1-27 Infantry Regiment "Wolfhounds." (Photo by SFC JB Jaso III)

CPT Christopher Hellie, of Duluth, MN, passes a rucksack to SPC Michael Shoaf, of Lewisville, NC, to be packed on a truck that transported the bags to the chartered civilian airplane at Kuwait City, Feb. 25, 2009. Hellie is the mortar platoon leader, while Shoaf is a mortarman with HHC/1-27 Infantry Regiment "Wolfhounds." (Photo by SFC JB Jaso III)

Soldiers from 1-27 Infantry Regiment "Wolfhounds" are greeted by military leaders, including LTG Benjamin Mixon, after arriving at Hickam Air Force Base in Honolulu, HI on Feb. 21, 2009. Mixon is the Commanding General for United States Army, Pacific. (Courtesy Photo)

Soldiers from HHC/1-27 Infantry Regiment "Wolfhounds" deplane the aircraft at Hickam Air Force Base in Honolulu, HI, following approximately 24-hours of travel from Kuwait City to Hawaii, Feb. 21, 2009. (Photo by SGT Brad Willeford)

Soldiers from 1-27 Infantry Regiment "Wolfhounds" sing the "Tropic Lightning March" during the welcome home ceremony held at Wheeler Army Airfield on Feb. 21, 2009. (Courtesy Photo)

The family of SPC Liam Norton, of Middletown, NJ, await to see their soldier at the welcome home ceremony held at Wheeler Army Airfield on Feb. 21, 2009. Norton is a reconnaissance scout with HHC/1-27 Infantry Regiment "Wolfhounds." (Photo by SGT Brad Willeford)

SPC Derek Martin, of Ironton, OH, grabs a snack and drink while waiting for the welcome home ceremony to begin at Wheeler Army Airfield on Feb. 26, 2009. Martin is an infantryman with A/1-27 Infantry Regiment "Wolfhounds." (Photo by MAJ Al Hing)

SSG Ramiro Ascencio, of Pomona, CA, reads the serial number of his M4 carbine to SPC Blake Kerr, of Bonnots Mill, MO, before turning it into the arms room, Feb. 26, 2009 at Wheeler Army Airfield, HI. Ascencio is a reconnaissance scout with HHC/1-27 Infantry Regiment "Wolfhounds." (Photo by MAJ Al Hing)

CPT Derek Pottinger, of Port Byron, IL, says a prayer prior to the welcome home ceremony at Wheeler Army Airfield on Feb. 26, 2009. Pottinger is the chaplain for 1-27 Infantry Regiment "Wolfhounds." (Photo by MAJ Al Hing)

Soldiers from 1-27 Infantry Regiment "Wolfhounds" sing the "Army Song" during the welcome home ceremony held at Wheeler Army Airfield on Feb. 26, 2009. (Photo by MAJ Al Hing)

WELCOME HOME WOLFHOUNDS
Welcome Home
WELCOME HOME
Mahalo for a Job Well Done!
MWR

WELCOME
HOME
DADDY

LTC Richard "Flip" Wilson, of Brooklyn, NY; and CSM William Hain, of Reno, NV uncase the battalion colors during a ceremony held at Sills Field, Schofield Barracks, HI, on March 12, 2009. The battalion's colors were cased on Feb. 22, 2009 at Camp Taji, Iraq, officially ending the 15-month deployment. (Photo by SFC JB Jaso III)

Jennifer Santistevan is happy to welcome back her boyfriend, SGT Michael Rios, of Sebastian, TX, after a welcome home ceremony held at Sills Field on Schofield Barracks, HI, March 12, 2009. Rios is an infantry team leader with A/1-27 Infantry Regiment "Wolfhounds."
(Photo by SFC JB Jaso III)

Chelsea Miller welcomes back her husband, SPC Stephen Miller, of Batesville, AR, following the 2nd Stryker Brigade Combat Team welcome home ceremony held on March 12, 2009 at Schofield Barracks, HI. Miller is an infantryman with A/1-27 Infantry Regiment "Wolfhounds."
(Photo by SFC JB Jaso III)

Sarah Murphy joins her husband, SGT Steven Murphy, of Boston, on Sills Field following the 2nd Stryker Brigade Combat Team welcome home ceremony at Schofield Barracks, HI on Mar. 12, 2009. Murphy is an infantryman with B/1-27 Infantry Regiment "Wolfhounds."
(Photo by SFC JB Jaso III)

Jenni Smith welcomes back here husband, SGT Vern Smith, of Channahon, IL, following the 2nd Stryker Brigade Combat Team welcome home ceremony at Schofield Barracks, HI on Mar. 12, 2009. Smith is a mortarman with A/1-27 Infantry Regiment "Wolfhounds."
(Photo by SFC JB Jaso III)

SPC Steven Beal, of Lavergne, TN, holds his daughter who is so excited to see Daddy again following the 2nd Stryker Brigade Combat Team welcome home ceremony at Schofield Barracks, HI on Mar. 12, 2009. Beal is an infantryman with B/1-27 Infantry Regiment "Wolfhounds."
(Photo by SFC JB Jaso III)

SFC Russell Stacy informs newly arriving Soldiers regarding the policies and regulations of U.S Army Hawaii at Schofield Barracks. Stacy is the rear detachment NCO for 1-27 Infantry Regiment "Wolfhounds." (Photo by SSG JB Jaso III)

Rear Detachment Soldiers for 1-27 Infantry Regiment "Wolfhounds." (Photo by SSG JB Jaso III)

The 1-27 Infantry Regiment "Wolfhounds" Rear Detachment conducted operations from the "Wolfhounds" headquarters at Schofield Barracks, where they assisted families allowing the forward deployed Soldiers to remain operationally focused. (Photo by SSG JB Jaso III)

MAJ Cory Marlowe, Rear Detachment OIC; and SFC Russell Stacy, Rear Detachment NCO for 1-27 Infantry Regiment "Wolfhounds." (Photo by SSG JB Jaso III)

SSG Douglas Pearson reviews administrative requirements that must be accomplished prior to the return of soldiers deployed in Iraq, Dec. 9, 2008 at Schofield Barracks. Pearson is the rear detachment administration NCO for 1-27 Infantry Regiment "Wolfhounds."
(Photo by SSG JB Jaso III)

MAJ Cory Marlowe responds to e-mail regarding redeployment operations at Schofield Barracks. Marlowe is the rear detachment OIC for 1-27 Infantry Regiment "Wolfhounds."
(Photo by SSG JB Jaso III)

Ms. Erica Gantt assists in planning, coordinating, and executing of family support services for family members separated due to the deployment from Schofield Barracks. Gantt is the Family Readiness Support Assistant for 1-27 Infantry Regiment "Wolfhounds."
(Photo by SSG JB Jaso III)

MAJ Cory Marlowe discusses the plans for the welcome home festivities and celebrations with members of the battalion's rear detachment on Dec. 9, 2008 at Schofield Barracks. Marlowe is the rear detachment OIC for 1-27 Infantry Regiment "Wolfhounds."
(Photo by SSG JB Jaso III)

Members of the 1-27 Infantry Regiment's Family Readiness Group that participated in the Camp Warrior Day program at Schofield Barracks, Hawaii.

Members of the Battalion's Family Readiness Group gather to make Halloween Bags for "Wolfhound" Soldiers deployed to Iraq.

Wolfhound children enjoy their day at Camp Warrior Day at Schofield Barracks, Hawaii

Members of the 1-27 Infantry Regiment's Family Readiness Group, along with the BN's Rear Detachment conducted several car washes to raise money for FRG activities and to send much needed supplies and goodies to the "Wolfhound" soldiers deployed.

Wolfhound "Pups" enjoying Camp Warrior Day activities at Schofield Barracks, Hawaii.

SFC Russell, Heather and Hailee Stacy during Camp Warrior Day festivities at Schofield Barracks. Russell is the 1-27 Infantry Regiment Rear Detachment NCOIC, while Heather is HHC's FRG leader.

Members of the 1-27 Infantry Regiment's Family Readiness Group stuff Christmas stockings to send to the "Wolfhounds" deployed in Iraq for their second consecutive holiday season.

BN Command Sergeant Major & Battalion Commander

CSM William Hain (BN CSM); LTC Richard "Flip" Wilson (BN CDR) (Photo by SFC JB Jaso III)

BN Executive Officer & Operations Officer

MAJ Patrick Aspland (BN XO); MAJ Michael Zernickow (BN S3) (Photo by SFC JB Jaso III)

BN Operations Sergeant Major

SGM Timothy Foster (BN OPS SGM)
(Photo by SPC Brian Pierce)

S4 Shop

SSG Jose Cobian; SGT Jimmy Prince; CPT Chad Chenoweth (S4 OIC); SFC Jerry Cox (S4 NCO); SSG Richard Durgin
(Courtesy Photo)

ISF Cell

SPC Bush; CPT Omololu Makinde (ISF Coordinator); SPC David Podelco; SPC David Godwin (Photo by SPC Brian Pierce)

S1 Shop

Front Row: SSG Jeremie Olive; SGT Ulises Arellano; SFC James Rigdon (Section NCO); PVT Jose Algarin
Back Row: SSG Christopher Nelson; SPC Ishaun Byers; SGT Stephen Babcock; SPC Rickey Reed (Photo by SFC JB Jaso III)

embedded Local Reconstruction Team (eLRT)

CPT John Goldsworthy (Team Leader); 1LT John Busuego; SFC Jesse Okiyama; 1LT Erik Wiesehan (Photo by SPC Brian Pierce)

Unit Ministry Team

CPT Derek Pottinger (Chaplain); and SGT Aaron Tisdale (Chaplain's Assistant) (Photo by SPC Brian Pierce)

Command,
HHC/1-27 IN "Havoc"
1SG Timothy Jackson (First Sergeant); CPT Brian Sweigart (Commander) (Photo by SPC Brian Pierce)

JSS Istaqlal Mayor's Cell
SGT Robert Lacy; MSG Casey Bunnell (JSS Mayor); SSG Heath Henley (Photo by SGT Brad Willeford)

Company HQ,
HHC/1-27 IN "Havoc"
SGT Christopher Clay; SGT Benjamin Eckhardt; SSG Carlos Robles; SGT Raul Ventura (Photo by SFC JB Jaso III)

HQ, Reconnaissance Platoon, HHC/1-27 IN "Havoc"

CPT Peter Walther (Platoon Leader); SFC Benjamin Chapman (Platoon Sergeant) (Photo by SGT Brad Willeford)

X Team, Reconnaissance Platoon, HHC/1-27 IN "Havoc"

SPC Luke Butler; SGT Brandon Baricuatro; PFC Robert Hatala; SSG Michael Mucha (Team Leader); SPC Brandon Nelson; SGT Nathan Adams; SPC Osvaldo Ortiz (Photo by SGT Brad Willeford)

Y Team, Reconnaissance Platoon, HHC/1-27 IN "Havoc"

SPC Clinton Coey; SPC Andrew Schultz; SPC Ronald Jarrard; SSG James Hall (Team Leader); SPC Jacob Bruno (Platoon Medic) (Photo by SGT Brad Willeford)

Z Team, Reconnaissance Platoon, HHC/1-27 IN "Havoc"

SSG Ramiro Ascencio; SGT John Singleton; SPC Joel Wright; SPC Nathan Phillips; SPC Joshua Fulgham; SSG Mark O'Neill (Team Leader) (Photo by SGT Brad Willeford)

Sniper Section, HHC/1-27 IN "Havoc"

Front Row: PFC Daniel Olsen; PFC William Norton; SGT Timothy Hobbs

Back Row: SPC Michael Haenel; SSG Devin Dove (Section Leader); PFC David Newkirk; SGT Curtiss Leilich

(Photo by SFC JB Jaso III)

Command, Medical Platoon,
HHC/1-27 IN "Havoc"
MSG Casey Bunnel (Platoon Sergeant); 1LT Erik Wiesehan (Platoon Leader)

PSD-1,
HHC/1-27 IN "Havoc"
Front Row: SPC Gregory Kastner (Section RTO); SPC Joseph Biando; SGT Justin Wells; SPC Allen Post (Section Medic); SGT Daniel Van Houten
Back Row: SGT Christopher Tinkham; SPC Matthew Lowery; SGT Richard Bobbitt; CPT Christopher Hellie (Platoon Leader); SPC Gordon Jordan; SPC Christian Diaz; SSG Caleb Crosby (Section NCO) (Photo by SFC JB Jaso III)

PSD-2,
HHC/1-27 IN "Havoc"
Front Row: SPC Joseph Kendall; PFC Wilfred Ferris; PFC Ishaun Byers
Back Row: CPT Christopher Hellie (Platoon Leader); SSG Kellen Hansen (Section NCO); SPC Rickey Reed; SPC Erich Young (Section Medic) (Photo by SFC JB Jaso III)

PSD-3,
HHC/1-27 IN "Havoc"
Front Row: SGT Emanuel Alvarez; SGT James Gompper; SPC Edgar Gutierrez
Back Row: SPC Michael Shoaf; CPT Brian Sweigart (Company Commander); SPC David Dempsey; SGT Eric Martin; SPC Timothy Russell; SPC Adam Bolderson; SSG Jesus Robles (Section NCO); SGT Kenneth Severin (Section Medic); SPC John Vernon; SPC Austin Sandoval; SGT Joshua Oden; SPC Brian Barnett; CPT Christopher Hellie (Platoon Leader); SGT Eric Waddle
(Photo by SGT Brad Willeford)

Medical Platoon,
HHC/1-27 IN "Havoc"
Front Row: Kolchak XIV
Middle Row: SPC Erich Young; PVT Ryan Maxfield; SGT Malaefou Heather; SPC William Groves; SPC Michael Kaufman
Back Row: SSG Vincent Valencia; SSG Jeffrey Coots; SGT Kenneth Severin; SGT Adam Costello; SPC Jacob Bruno; SGT Raul Ventura; SSG Jose Perez (Photo by SFC JB Jaso III)

Command, A/1-27 IN "Gunslingers"

CPT Lucas Yoho (Company Commander); 1SG Eugene Dydasco (First Sergeant) (Photo by SPC Brian Pierce)

CMD, Weapons Platoon, A/1-27 IN "Gunslingers"

SFC Dion Walker (Platoon Sergeant)
(Photo by SFC JB Jaso III)

Company Headquarters, A/1-27 IN "Gunslingers"

Front Row: SPC Michael Dean; SGT Raymond Hepworth; SGT Mark Davenport; SSG George Canfill
Back Row: SSG Michael Pharr; SPC Jacob Atchison; SSG Glenn Devitt; SSG Larry Duda; SPC Eric Gonzalez; SPC Nicholas Snyder; PFC Robert Alvarez
(Photo by SFC JB Jaso III)

Mortar Section, Weapons Platoon, A/1-27 IN "Gunslingers"

Front Row: SGT Christopher Vacarro; SPC Ryan Wilson; SPC Corey Show; SPC Brandon Paronett; PFC Brenton Baer
Back Row: SSG Joseph Simpliciano (Section Leader); SGT Vern Smith; SPC Brandon Schneider; SPC Yee Vang; SPC Cameron Thoreson; PFC Richard Spence
(Photo by SFC JB Jaso III)

MGS Section, Weapons Platoon, A/1-27 IN "Gunslingers"

Front Row: SGT Cody Bagwell; SPC Jehad Mahmoud; PFC Erik Christiansen; PFC Matthew Dunkel
Back Row: SSG Bryan Starr; SGT John Yokeley; SFC Dion Walker (Platoon Sergeant); SGT Robert Scott; SGT Bryan Sawyer
(Photo by SFC JB Jaso III)

CMD, 1st Platoon, A/1-27 IN "Gunslingers"

2LT Christopher Hair (Platoon Leader); SFC Stephen Novak (Platoon Sergeant) (Photo by SPC Brian Pierce)

1st Squad, 1st Platoon, A/1-27 IN "Gunslingers"

Front Row: SSG Tony Dalton; SPC Jose Nunez; PFC Matthew Aungst; SPC Nicholas Nystrom
Back Row: SGT David Exton; SGT James Simmons; SSG Michael Clark (Squad Leader); SPC Jonathan Mendez; SGT Bruce Dixon
(Photo by SFC JB Jaso III)

2nd Squad, 1st Platoon, A/1-27 IN "Gunslingers"

Front Row: SSG Jonathan Castro; SSG Dusty Lamborn; SPC James Childs
Back Row: PFC Victor Alaniz; SPC Corey Payne; PFC Marlon Esparraguera; SGT Michael McConaghy; SPC Richard Lyons; SGT Kenneth Larsen; SPC Brett Leback
(Photo by SPC Brian Pierce)

HQ, 1st Platoon, A/1-27 IN "Gunslingers"

SPC Adam McClure (Platoon Forward Observer); SPC Charles Joseph (Platoon RTO); SPC Thomas Bateman (Platoon Medic)
(Photo by SPC Brian Pierce)

3rd Squad, 1st Platoon, A/1-27 IN "Gunslingers"

Front Row: SPC James Robitaille Jr., SPC Ruben Salas; SPC Trey Wright
Back Row: SSG Jeremiah Harpster; SPC Matthew Ferguson; SPC Donald Mollis; SSG Roberto Zuniga (Squad Leader); SPC Jonathan Riggs (Photo by SPC Brian Pierce)

Weapons Squad, 1st Platoon, A/1-27 IN "Gunslingers"

CPL Zachary Manuel; SGT Cody Travis; SPC Richard Martinez; SPC Charles Joseph (Platoon RTO); SPC Kevin Butler; CPL Timothy Terwilliger; SPC Thomas Bateman (Platoon Medic); SPC Adam McClure (Platoon Forward Observer)
(Courtesy Photo)

CMD, 2nd Platoon, A/1-27 IN "Gunslingers"

SFC James Burciaga (Platoon Sergeant); 1LT Tommy Ryan (Platoon Leader) (Photo by SPC Brian Pierce)

1st Squad, 2nd Platoon, A/1-27 IN "Gunslingers"

SGT Bryon Davis; PFC Sean Cox; SGT Timothy Miller; PFC Justin Do; SGT Arthur Smith; SGT Michael Rios; SPC Scott Blazek; SGT Garrett Simoneau; SSG Christopher Bolwell (Squad Leader)
(Photo by SPC Brian Pierce)

2nd Squad, 2nd Platoon, A/1-27 IN "Gunslingers"

SPC Kyle Lauth; SPC Joshua Lewis; SSG Rodriquez Cruell (Squad Leader); SPC Christopher Mola; CPL Jeremy Mingle; SGT Pedro Macias; SGT Eberardo Solis; SPC David Minsink
(Photo by SPC Brian Pierce)

3rd Squad, 2nd Platoon, A/1-27 IN "Gunslingers"

SPC Corey Johnson; SSG Andrew Lambright; SSG Brian Wallace (Squad Leader); SGT Collin Kelly; SPC Stephen Miller; SGT Michael Miller; SGT Ilalio Saufoi
(Photo by SPC Brian Pierce)

Weapons Squad, 2nd Platoon, A/1-27 IN "Gunslingers"

SFC James Burciaga (Platoon Sergeant); SPC Samuel Fowler; SGT Charles Haygood; SPC Scott Weller; SPC Lance Moore; SPC Johnathan Read; SPC Joshua Lloyd; SPC Alonzo Martin; SGT Jerimiah Sullivan (Squad Leader); SGT Joseph Baker; SPC Joshua Echeverria; 1LT Tommy Ryan (Platoon Leader)
(Photo by SPC Brian Pierce)

CMD, 3rd Platoon, A/1-27 IN "Gunslingers"

SFC Omar Ashley (Platoon Sergeant); 1LT Christopher McCoy (Platoon Leader) (Photo by SFC JB Jaso III)

1st Squad, 3rd Platoon, A/1-27 IN "Gunslingers"

Front Row: SPC Timothy Moreland; SPC Benjamin Raulerson; PFC Richard Plasket; SPC Leonard Heiniger
Back Row: SPC Eric Wright; SGT Ryan Ruz; SSG Curtis Johnson (Squad Leader); SGT Robert Harryman
(Photo by SFC JB Jaso III)

2nd Squad, 3rd Platoon, A/1-27 IN "Gunslingers"

Front Row: SPC Jon Bothke; PFC Jeffrey Arnold; PFC Matthew McMahon
Back Row: SGT Abraham Norton; SSG Louis Overson (Squad Leader); SGT Gary Fox (Photo by SFC JB Jaso III)

HQ, 3rd Platoon, A/1-27 IN "Gunslingers"

SGT Dustin Lapolla (Platoon Forward Observer); and SGT Benjamin Dang (Platoon Medic) (Photo by SFC JB Jaso III)

3rd Squad, 3rd Platoon, A/1-27 IN "Gunslingers"

Front Row: SPC Jaymes Beebe; PFC Kenneth Bledsoe; SPC John Mitchell; SPC Jesse Weber
Back Row: SPC Samuel Harris; CPL Eugene Dimauro; SSG Daniel Ramirez (Squad Leader); SGT Douglas Urias
(Photo by SFC JB Jaso III)

Weapons Squad, 3rd Platoon, A/1-27 IN "Gunslingers"

Front Row: SPC Roland Manalastas; SPC Steven Williams; SPC Christopher Jones; SPC David Denson
Back Row: SPC Julio Lopez; SPC Jason Judd; SGT Nathan Taylor (Squad Leader); SGT Napoleon Gonzalez
(Photo by SFC JB Jaso III)

Command, B/1-27 IN "Borzoi"

CPT Andrew Farina (Company Commander); 1SG Todd Ballenger (First Sergeant)
(Photo by SPC Brian Pierce)

CMD, Weapons Platoon, B/1-27 IN "Borzoi"

1LT Louis Cascino (Platoon Leader); SFC Justin McCrae (Platoon Sergeant)
(Photo by SGT Brad Willeford)

Company Headquarters, B/1-27 IN "Borzoi"

Front Row: SPC Erik Ramon; SSG Andre Cornwell; SPC Kenneth McKinney; SPC Jeremiah Robles; SPC Michael Waldie. Back Row: SSG John White; CPT Sean Hollars (Executive Officer); SSG Jeffrey Davis; SGT Daniel York; SGT Jared Gabriel; SSG Bruce Hillier; SGT Jason Stechman; SGT Patrick Adams; SSG Eric Mulkey; SPC Darrin Bush; SGT Luis Barrionuevo; SGT Matthew Bair; SSG Brian Murphy (Platoon Sergeant)

Mortar Section, Weapons Platoon, B/1-27 IN "Borzoi"

Front Row: SPC Drew Stegehuis; CPL Michael Tejeda; SPC Jesse Arnold; SPC Darryn Walton
Back Row: SSG Kelly Muna (Section Leader); SPC Timothy Larose; SPC Joshua Foster; SPC David Sorensen; PFC Ryan Winkelmann; SGT Eric Burke
(Photo by SGT Brad Willeford)

MGS Section, Weapons Platoon, B/1-27 IN "Borzoi"

Front Row: SSG Rudy Mello (Section Leader); SGT Ian Bittenbender; SGT Edward Mara; SGT Joselito Batocabe; PFC Dwayne Haynes (Platoon Medic); SPC Shaun Dorfe
Back Row: SPC Bryan Landers; SPC Clifford Martin; SGT Andrew Holland; SPC Eric Barcomb; SPC Gilbert Sandoval; SPC Douglas Byers; SPC Eric Esquivel
(Photo by SGT Brad Willeford)

CMD, 1st Platoon, B/1-27 IN "Borzoi"

CPT Richard Pace (Platoon Leader); SFC Jasmer Moses (Platoon Sergeant)
(Photo by SPC Brian Pierce)

1st Squad, 1st Platoon, B/1-27 IN "Borzoi"

Front Row: SPC Kent Smith; SPC Parker Smith; SPC Albert Peralta
Back Row: SPC Andrew Hart; SGT Cameron Carney; SSG Rober Hirschi (Squad Leader); SGT Mason Roy; SGT Sebastian Klui
(Photo by SPC Brian Pierce)

2nd Squad, 1st Platoon, B/1-27 IN "Borzoi"

Front Row: SPC Matthew Slagle; SGT Jackie Smith; SGT Pius Hadley
Back Row: PFC Daniel Perryman; SSG Jeremy Beer; SSG Michael Ortiz (Squad Leader); SGT Matthew Baker; SPC David Podelco; SPC Brook Lindsey (Photo by SPC Brian Pierce)

HQ, 1st Platoon, B/1-27 IN "Borzoi"

SGT Steven Murphy (Platoon RTO); SPC Joseph Lorenz (Platoon Medic); SGT William Berendt (Platoon Forward Observer)
(Photo by SPC Brian Pierce)

3rd Squad, 1st Platoon, B/1-27 IN "Borzoi"

Front Row: SSG Ernesto Rodriguez (Squad Leader); SPC Alan Bradford; SGT Travis Stegemann; SGT Jacob Purdy
Back Row: SPC Matthew Deagle; SPC Steven Beal; SPC Michael Reed; SGT Eteru Tafua (Photo by SPC Brian Pierce)

Weapons Squad, 1st Platoon, B/1-27 IN "Borzoi"

Front Row: SGT Steven Murphy (Platoon RTO); SGT William Holley; SPC Jonathan Stryker; SPC Kenneth Smith; SSG Patrick Santos (Squad Leader)
Back Row: PFC Aaron Paul; SSG James Peptis; SPC Dustin Smith; SGT William Berendt (Platoon Forward Observer); SPC Joseph Lorenz (Platoon Medic) (Photo by SPC Brian Pierce)

CMD, 2nd Platoon,
B/1-27 IN "Borzoi"

1LT Kenneth Jura (Platoon Leader); SFC Kristopher Verrett (Platoon Sergeant)
(Photo by SPC Brian Pierce)

1st Squad, 2nd Platoon,
B/1-27 IN "Borzoi"

Front Row: SGT Sean Cheker; SPC Whinny Jonas; SPC Eric Becker
Back Row: SSG David Fischgrund (Squad Leader); PFC Dillon Jayne; SPC Franklin Walz; SSG Jason Stalnaker
(Photo by SPC Brian Pierce)

2nd Squad, 2nd Platoon,
B/1-27 IN "Borzoi"

Front Row: SPC Jason Ellis; SPC Joshua Collins; PFC Billy Hill; SPC Tyler Royle
Back Row: SSG Brian Smith (Squad Leader); SGT Joseph Galvan; SPC Anthony Stevens; PFC Scott Paugel
(Photo by SPC Brian Pierce)

3rd Squad, 2nd Platoon,
B/1-27 IN "Borzoi"

Front Row: SGT Joseph Miramontez; PVT Joshua Castleman; SGT Edwin Brocaille
Back Row: PFC Kevin Higgins; PVT Christopher McGinnis; SGT James Craig (Squad Leader); SPC Jose Morales
(Photo by SPC Brian Pierce)

Weapons Squad, 2nd Platoon,
B/1-27 IN "Borzoi"

Front Row: SPC Zachary Gardner; SPC Randel Marcotte (Platoon Medic); PFC Derrick Kozlowski; SGT William Heuisler
Back Row: SPC Joshua Cole; SPC Zachary Singer; SSG Matthew Weitz (Squad Leader); SGT Christopher Watson (Platoon Forward Observer); SPC Jerad Nydam
(Photo by SPC Brian Pierce)

CMD, 3rd Platoon, B/1-27 IN "Borzoi"
SFC Carlito Dizon (Platoon Sergeant); 1LT Benjamin Flores (Platoon Sergeant)
(Photo by SPC Brian Pierce)

1st Squad, 3rd Platoon, B/1-27 IN "Borzoi"
Front Row: PFC Jesse Huckleberry; SPC Christopher Odell
Back Row: SPC William Gentry; SSG Wes Liberty; SSG Wiley Baker; SSG Raymond Diaz (Squad Leader); SGT Shane Kraeger; SGT Tomas Rodriguez; SPC Jason Goss
(Photo by SPC Brian Pierce)

2nd Squad, 3rd Platoon, B/1-27 IN "Borzoi"
SSG Joseph Dearth (Squad Leader); SGT Steven Price; SGT Frank Nauta; SPC Brett Seaver; SPC Daniel Cortes
(Photo by SPC Brian Pierce)

3rd Squad, 3rd Platoon, B/1-27 IN "Borzoi"
Front Row: SPC Timothy Brown; SPC James Arnett; SPC Derick Hannah;
Back Row: SGT Thomas Warmoth; SGT Adam Brown; SSG Charles Bentley (Squad Leader); SSG Denny Toussaint; SSG Philip Snook; SGT Jonathan Martin
(Photo by SPC Brian Pierce)

Weapons Squad, 3rd Platoon, B/1-27 IN "Borzoi"
Front Row: SGT Wesley Shipman; SPC Kirill Grinchenko; SGT Adam Long; SGT Aaron Robinson; SGT Alvino Sandoval
Back Row: SPC Christopher Eastman; SPC Wayne Cloud, Jr. (Platoon Forward Observer); SPC David Smith; SSG Lucas Collins (Squad Leader); SGT Kristopher Weber (Platoon Medic)
(Photo by SPC Brian Pierce)

Command, C/1-27 IN "Coldsteel"

CPT David Uthlaut (Company Commander); 1SG Robert Szymaszek (First Sergeant) (Photo by SFC JB Jaso III)

CMD, Weapons Platoon, C/1-27 IN "Coldsteel"

SFC Carl Holst (Platoon Sergeant); CPT Steven Robinson (Platoon Leader); SFC Kenneth Palumbo (MGS Section Leader) (Photo by SFC JB Jaso III)

Company Headquarters, C/1-27 IN "Coldsteel"

Front Row: SPC Anthony Cala; SGT James Hatcher; SGT Jessie Johnson; SPC Slagelferdy Oyao; SPC Thomas St. Maur; SPC Dervi Osorio; SPC Michael Vermillion; Back Row: SSG Jason Menard; SGT Aaron Miller; SGT Robert Pegg ;SGT Jonathan Pass; SGT Joseph Dunne; SFC Adam Lane (Platoon Sergeant); CPT Devon McRainey (Executive Officer); SPC Timothy Delgado; SGT Jesse Laenen; SSG Nigel James; SPC Robert Duffey; SSG Juan Sanson (Photo by SFC JB Jaso III)

Mortar Section, Weapons Platoon, C/1-27 IN "Coldsteel"

Front Row: SPC Michael Skirkanich (Platoon Medic); SPC Daniel Fulk; SPC Timothy Lewis; SPC Jeffrey Lavigne

Back Row: SPC Michael Vermillion; CPL Dustin Dewoody; SSG Nigel James (Section Leader); SPC Paul Pieritz; SPC Mark Shinn (Photo by SFC JB Jaso III)

MGS Section, Weapons Platoon, C/1-27 IN "Coldsteel"

Front Row: SPC Shannon Robinson; SGT Karl Schell; SGT Eric Wright

Back Row: SSG Afshin Nabavi; SGT Darryl Outlaw; SFC Kenneth Palumbo (Section Leader); SSG Timothy Williams; SPC Gregorio Torres (Photo by SFC JB Jaso III)

Command, 1st Platoon, C/1-27 IN "Coldsteel"

SFC Hugh Jones (Platoon Sergeant); 1LT Zachary Seidel (Platoon Sergeant) (Photo by SPC Brian Pierce)

1st Squad, 1st Platoon, C/1-27 IN "Coldsteel"

Front Row: SPC Arthur Guerra; SPC Christopher Bowman; PFC Daniel Ramirez; SPC Jeffrey Helon; SPC Jesse Youngstrom
Back Row: SGT Adam Lewis; SPC Nicholas Pata; SSG Jacey Callahan (Squad Leader); SGT Randy Stephens; SPC Shawn Crenshaw
(Photo by SFC JB Jaso III)

2nd Squad, 1st Platoon, C/1-27 IN "Coldsteel"

PFC Joshua Sims; SPC Larry Robertson; SGT Javier Snider; SSG Robert Klein (Squad Leader); SGT Jorge Gonzalez; SGT John Finn; SPC Paul Groth; SGT Vincent Vinciguerra
(Photo by SPC Brian Pierce)

HQ, 1st Platoon, C/1-27 IN "Coldsteel"

CPL Steven Vasko (Platoon Medic); SSG Jonathan Mhoon (Platoon Forward Observer) (Photo by SPC Brian Pierce)

3rd Squad, 1st Platoon, C/1-27 IN "Coldsteel"

Front Row: SPC Luis Figueroa; SSG Michael Blaisdell; SGT James Jeffcoat; SPC Michael Thao; SPC Douglas Kirk
Back Row: PFC Johnny Steele; SSG Nicholas Maxwell; SSG Mercedes Amaro (Squad Leader); SGT Andrew Churchill; SPC Daniel Lynch (Photo by SPC Brian Pierce)

Weapons Squad, 1st Platoon, C/1-27 IN "Coldsteel"

Front Row: PVT Rey Bartolome; SGT Joel Melin; SPC Shawn Crenshaw
Back Row: SSG Antonio Trillo (Squad Leader); SGT Aleksandr Litovchenko; SPC Austin Raske; CPL Steven Vasko; PFC Reed Stanley (Photo by SPC Brian Pierce)

CMD, 2nd Platoon, C/1-27 IN "Coldsteel"
1LT Marco Montoya (Platoon Leader); SFC Victor Benavides (Platoon Sergeant) (Photo by SFC JB Jaso III)

1st Squad, 2nd Platoon, C/1-27 IN "Coldsteel"
Front Row: SGT Daniel Mitchell; SPC David Maness; PFC David Paladini; SPC Toribio Gurrola
Back Row: PFC Kevin Bridges; SGT Ira Parenteau; SSG Israel Martinez (Squad Leader); SPC Travis Clabaugh; SGT Stuart Early
(Photo by SFC JB Jaso III)

2nd Squad, 2nd Platoon, C/1-27 IN "Coldsteel"
Front Row: SGT Wilburtigran Rayala; SGT Jacob Crowder; SPC William Ivonen; PFC Damian Ross
Back Row: SPC Christopher Sahu; SPC Michael Tull; SSG Lorne Wilson (Squad Leader); SPC Michael Tatum; SSG Robert Points
(Photo by SFC JB Jaso III)

3rd Squad, 2nd Platoon, C/1-27 IN "Coldsteel"
Front Row: SPC Daniel Clark; SPC Adam Scheid; SPC Jose Maganda
Back Row: SPC Edward Brodzinski; SGT Michael Stewart; SSG Irvin Toussaint (Squad Leader); SGT Broam Hurst
(Photo by SFC JB Jaso III)

Weapons Squad, 2nd Platoon, C/1-27 IN "Coldsteel"
Front Row: PFC Derek Smith; SPC Joshua Mears; SPC Matthew Dortch (Platoon Medic); SGT Mitchell Hassler (Platoon Forward Observer)
Back Row: SPC Bradley Cook; SSG Stephen Alejandro (Squad Leader); SGT Keith Washington; SPC Randall Schonfeld
(Photo by SFC JB Jaso III)

CMD, 3rd Platoon, C/1-27 IN "Coldsteel"
CPT Nicholas Pines (Platoon Leader); SFC Jesse Wyant (Platoon Sergeant) (Photo by SFC JB Jaso III)

1st Squad, 3rd Platoon, C/1-27 IN "Coldsteel"
Front Row: SPC Thomas Snowden; SGT Robert Haswell; PFC Katsone Phonevilay; and PFC Benjamin Garcia
Middle Row: SGT Donald Mathie; SGT Steven Thompson; SGT Thermond Robinson; and CPL Aaron Young
Back Row: SSG Joseph Stalinksi (Squad Leader)
(Photo by SFC JB Jaso III)

2nd Squad, 3rd Platoon, C/1-27 IN "Coldsteel"
Front Row: PFC Yann Paquet; SPC Phillip Cody; SGT Steven Walters; PFC David Stretch; and SGT Shea Thames
Middle Row: PFC Nicholas Dahl; SPC Adam Riddle; SGT Antwan Scales; and PFC Jonathan Pendelton
Back Row: SSG James McCall (Squad Leader)
(Photo by SFC JB Jaso III)

HQ, 3rd Platoon, C/1-27 IN "Coldsteel"
PFC Brandon Pillen (Platoon RTO); CPL Jeffrey Niesluchowski (Platoon Medic) (Photo by SFC JB Jaso III)

3rd Squad, 3rd Platoon, C/1-27 IN "Coldsteel"
Front Row: SPC Jeremy Picinic; SPC Phillip Cody; PFC Chance Shrader
Middle Row: SGT Nicholas Spottedbear; SGT Brian Cochran; SSG Benjamin Curneal; SPC Michael Jacobs
Back Row: SSG Michael McMurray (Squad Leader)
(Photo by SFC JB Jaso III)

Weapons Squad, 3rd Platoon, C/1-27 IN "Coldsteel"
Front Row: PFC William Washburn; SGT Karl Rhea; PFC Brian Renzi
Middle Row: SGT David Lopez; SPC Seth Wood; SGT Cocaine Perry
Back Row: SSG James Walker; SSG Harry Jenkins (Squad Leader)
(Photo by SFC JB Jaso III)

CMD, 1st Platoon, B/52 IN "Hammer"

SFC Shane Cleghon (Platoon Sergeant); SPC John Ward (Platoon Medic); 1LT Joshua Kordela (Platoon Leader)
(Photo by SFC JB Jaso III)

CMD, B/52 IN "Hammer"

1SG Shiloh Butterworth (First Sergeant) and CPT Darrell Tran(Company Commander) (Photo by SPC Brian Pierce)

CMD, 3rd Platoon, B/52 IN "Hammer"

1LT Eddie Thomas (Platoon Leader) and SFC Raymond Zwicker (Platoon Sergeant) (Photo by SPC Brian Pierce)

2nd Platoon, B/52 IN "Hammer"
Front Row: SPC Travis Craig; SPC Alexignacio Trinidad; SPC Jesse Barnes; PFC Aaron Fenn; PFC Dwayne Haynes (Platoon Medic); SPC Clifford Lasure
Back Row: SSG Gregory Antoine; SGT Scott Geiszler; SSG Joseph Perminas (Platoon Sergeant); SGT George Davis; SSG David Ni; SPC Bryan Landers; (Photo by SGT Brad Willeford)

Company Headquarters, B/52 IN "Hammer"
Front Row: CPL James Reed; SPC Jonathan Barnett; SGT James Gutierrez
Back Row: CPT Will Lincoln (Company XO); SGT Donnell Macklin; SSG Joseph Brown; SSG Thomas Goodwin; SSG Christopher Miklas (Photo by SPC Brian Pierce)

3rd Platoon, B/52 IN "Hammer"
Front Row: SSG Brandon Tulloch; SPC Nicholas Arruda; SPC Douglas Graham; SPC James Robitaille; SPC Riki Tate; SPC Joseph Santa Lucia
Back Row: SSG Nathan Delander; SGT Justin Bradish; SPC William Jackson; SPC Alexis Castaneda; SSG Ruben Martinez, Jr.
(Photo by SPC Brian Pierce)

Tactical Psychological Operations Team (TPT) 1624

SPC Jarred Carruthers, SSG Donald Fabian (Team Leader), and SPC John Pearson (Photo by SFC JB Jaso III)

Tactical Psychological Operations Team (TPT) 1231

SPC David Bush, SGT Ryan Davis (Team Leader), and CPL Steven Randall (Photo by SFC JB Jaso III)

Tactical Psychological Operations Team (TPT) 1212

SPC Jared Hollingsworth, SSG Charlie Sanchez (Team Leader), and SPC John Doss (Photo by SPC Brian Pierce)

Command Team, Combat Repair Team

SSG Joshua Lang; CW3 Danny Taylor (Maintenance Technician); SFC Charles Truesdell (BN Motor Sergeant)
(Photo by SPC Brian Pierce)

Camp Taji, Combat Repair Team

Front Row: CW3 Danny Taylor; SPC Sean Olsen; SGT Emeluterio Lopez; SPC Donald Ross; SPC Jose Carrillo; SFC Charles Truesdell

Back Row: SPC Tajalli Hannibal; SGT David Wilson; SGT Kevin Green; SSG Joshua Lang; SGT Nicolas Laduke; SPC Jermaine Anderson; SPC Thomas McPherson; SPC Daniel Stevens; SPC Aubrey Schrader; SGT Matthew Keeling; SPC Nathanial Eckhoff
(Photo by SPC Brian Pierce)

Camp Taji, General Dynamics Land Systems, Combat Repair Team

Erica Corral; Juan Alvarez; Nathan Bright; Tommy Jackson; Carlito De Leon; Amor Lubina
(Photo by SPC Brian Pierce)

STG
CTR2 Matthew Todd; CTR1 Scott Lamb
(Photo by SFC JB Jaso III)

Electronics Warfare Officers
CTR1(SW) Nathan Francis; SSG Heath Henley
(Photo by SFC JB Jaso III)

JSS Hor al Bosh Cooks
SGT Jerud Udelhofen; SSG Jonathan Castro
(Photo by SFC JB Jaso III)

JSS Istaqlal, General Dynamics Land Systems, Combat Repair Team
Emmett Lozano; Kevin Koehler
(Photo by SGT Brad Willeford)

JSS Istaqlal, Combat Repair Team
SSG Mitchell Saltsman; SGT Berry Lebehn; SPC Lee Coddington; SPC Nathanial Eckhoff (Photo by SPC Brian Pierce)

Laundry Team, 590th QM Co., 548th CSSB
PFC Karla Chimbaquedano; SPC Kathleen Stumpner; SPC Jose Gomez; SGT Armando Soriano (Photo by SGT Brad Willeford)

JSS Istaqlal Cooks
SPC Kristopher Curtis; SSG Christopher Dent; SGT James Snow; SPC Christopher Benton; SSG Rizalino Corpuz
(Photo by SGT Brad Willeford)

Soldiers from 1-27 Infantry Regiment "Wolfhounds" are formed on Sills Field, Schofield Barracks, HI, during the 2nd Stryker Brigade Combat Team "Warrior" welcome home ceremony March 12, 2009, officially ending a 15-month deployment to Iraq.
(Photo by SFC JB Jaso III)

Acknowledgments

I have closely watched the work of hundreds of reporters and photographers, both civilian and military, since the beginning of Operation Iraqi Freedom. Never have I seen work that is as professional and top quality as the work produced by the 1-27 IN "Wolfhounds" Audio/Visual team of SFC JB Jaso, III; SGT Brad Willeford; SPC Brian Pierce; and SPC John Ahn. Long before I knew about the interest in producing this book, I was in awe of their work, often done under the risk of enemy attack. Without this top notch team of photographers, this book would not have been possible. We should all thank this team of professionals who cared enough to preserve these memories for future generations of Americans to enjoy and learn from. We also thank the other photographers who contributed.

CPT Brian Sweigart deserves special thanks. As he returned to Iraq from R&R, it was a chance meeting at the Atlanta airport USO where we talked about the vision of a "Wolfhound Reflections" book. My work as a USO volunteer, with the Veterans History Project to preserve memories for future generations, and in supporting the Soldiers and Families of MND-B, coupled with my retirement job as an author/publisher, made this book idea jump out at me as something I wanted to help make into a reality. Brian Sweigart planted that seed. Without him, it's unlikely a publisher who would donate his time, money, and effort would have been found.

SFC JB Jaso III worked tirelessly, both in Iraq and giving up his leave upon returning to Hawaii, to make this book happen. Decisions on what to include and exclude, insuring accuracy, shooting more pictures to fill holes, and driving early sales fell solely on JB Jaso's shoulders – jobs he did flawlessly. No single person was more important to this book than was JB Jaso.

No book becomes a reality without the hard operational work of editing; designing and laying out the pages; making untold changes as we fine tuned to make it the best it could be; and balancing the deadline pressures of insuring we included the final welcome home ceremonies with the reality of getting the book printed quickly. Mark Babcock was the man who made all that happen. He has a high respect for America's Soldiers and took pride in making this into a book that will be treasured by past, present, and future generations of Wolfhounds and their Families.

All books have costs associated with them. Those who pre-ordered, betting their money on a project that had just begun, can't be thanked enough. As more money was needed, a plea was sent out by SFC Jaso and Bob Babcock to their network of friends. The response from patriotic Americans was nothing short of phenomenal. Donations of all sizes came from Family, friends – IBM friends, church friends, high school friends, neighbors, and others – veterans and non-veterans, including some we don't know exactly how they learned about the need. To all of you, our heartfelt thanks – you are truly great, patriotic Americans.

Finally, the vision and commitment to preserving the history of the Wolfhounds for future generations came from LTC Richard "Flip" Wilson, the 1-27 IN battalion commander. LTC Wilson was unwavering in his desire to get this book produced for his Soldiers, their Families, and past and future generations of "Wolfhounds". Hopefully, this will be the standard that future generations of "Wolfhounds", as well as other units, will strive to follow.

As a fellow infantryman (Vietnam platoon leader with B/1-22 IN), I am proud to have had the honor of working with all those mentioned above in creating this book. Paraphrasing what George C. Scott said in the movie, "Patton" – when your grandchild climbs onto your lap and asks what you did during the war in Iraq, you can show him or her that you did your part.

Bob Babcock, Publisher – Deeds Publishing

Mark Babcock, Bob Babcock, and SFC JB Jaso III met at Bob's home in Marietta, GA to discuss *Wolfhound Reflections.*